AS UNIT 2

STUDENT GUIDE

CCEA

Chemistry

Further physical and inorganic chemistry and an introduction to organic chemistry

Alyn G. McFarland

HODDER
EDUCATION
AN HACHETTE UK COMPANY

Hodder Education, an Hachette UK company, Blenheim Court, George Street, Banbury, Oxfordshire OX16 5BH

Orders

Bookpoint Ltd, 130 Park Drive, Milton Park, Abingdon, Oxfordshire OX14 4SE

tel: 01235 827827

fax: 01235 400401

e-mail: education@bookpoint.co.uk

Lines are open 9.00 a.m.–5.00 p.m., Monday to Saturday, with a 24-hour message answering service. You can also order through the Hodder Education website: www.hoddereducation.co.uk

© Alyn McFarland 2016

ISBN 978-1-4718-6397-4

First printed 2016

Impression number 5 4 3 2 1

Year 2020 2019 2018 2017 2016

This guide has been written specifically to support students preparing for the CCEA AS and A-level Chemistry examinations. The content has been neither approved nor endorsed by CCEA and remains the sole responsibility of the author.

Cover photo Iamax/Fotolia

Typeset by Integra Software Services Pvt. Ltd, Pondicherry, India

Printed in Italy

Hachette UK's policy is to use papers that are natural, renewable and recyclable products and made from wood grown in sustainable forests. The logging and manufacturing processes are expected to conform to the environmental regulations of the country of origin.

Contents

Getting the most from this book 4
About this book 5

Content Guidance

Formulae and amounts of substance 6
Nomenclature and isomerism in organic compounds 14
Alkanes 33
Alkenes 38
Halogenoalkanes 46
Alcohols 57
AS organic identification tests 62
Infrared spectroscopy 63
Energetics 65
Kinetics 77
Equilibrium 80
Group II elements and their compounds 86

Questions & Answers

Formulae and amounts of substance 94
Alkanes and alkenes 96
Halogenoalkanes 99
Alcohols and infrared spectroscopy 101
Energetics 103
Kinetics and equilibrium 105
Group II elements and their compounds 108

Knowledge check answers 110
Index 111

■ Getting the most from this book

Exam-style questions

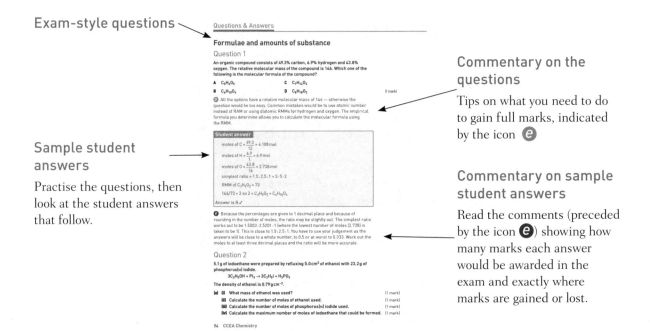

Commentary on the questions

Tips on what you need to do to gain full marks, indicated by the icon **e**

Sample student answers

Practise the questions, then look at the student answers that follow.

Commentary on sample student answers

Read the comments (preceded by the icon **e**) showing how many marks each answer would be awarded in the exam and exactly where marks are gained or lost.

■ About this book

This book will guide you through CCEA Chemistry AS Unit 2.

The **Content Guidance** covers the content of AS2 unit and gives helpful tips on how to approach revision and improve exam technique. Do not skim over these tips — they provide important guidance. There are also knowledge check questions throughout this section, with answers at the back of the book. At the end of each section there is a summary of the key points covered. There are several areas in this unit that are essential to other units in the course — for example, nomenclature and isomerism, energetics, equilibrium and kinetics. These topics and indeed all the content of this unit can be examined synoptically in A2 units.

The **Questions & Answers** section gives sample examination questions on each topic as well as worked answers and comments on the common pitfalls to avoid. The examination will consist of ten multiple-choice questions (each with four options, A to D), followed by several structured questions. This section contains many different examples of questions, but you should also refer to past papers for this unit, which are available online.

The Content Guidance and the Questions & Answers section are divided into the topics outlined by the CCEA specification.

General tips

When answering questions involving the colour of a chemical, you must be accurate to obtain the marks. If the colour of a chemical is given in this book with a hyphen between the colours then state the two colours exactly like that, including the hyphen. For example, the Cu^{2+} flame test colour is blue-green so both blue and green are required, separated by a hyphen. If two or more colours are given separated by a forward slash (/), these are alternative answers and only **one** colour from this list should be given. For example, bromine water is described as yellow/orange/brown so only yellow on its own or orange on its own or brown on its own will be accepted but **not** a combination of the colours. **Never use a forward slash when answering a colour question.** If only one colour is given for a chemical then use this single colour — for example, orange for the colour of acidified potassium dichromate(vi) solution. This applies to all CCEA AS and A2 examinations. Check the acceptable colours document on the CCEA chemistry website (www.ccea.org.uk/chemistry) then select revised GCE for further guidance should this change.

Give the exact detail given in the acceptable colours document, any practical documents and the clarification of terms document (published on the CCEA website) to ensure you are awarded the marks. One word used incorrectly could lose you a mark. Be careful with particles terms such as atom, ion, molecule, radical, as these can cost you marks. Always give the correct positive and negative signs for charges (e.g. 2+) and oxidation states (e.g. +2) where appropriate.

20% of the marks at A-level are for mathematical skills.

Content Guidance

Formulae and amounts of substance

Empirical formula

The empirical formula of a compound is the simplest formula that states the composition of the compound. The **empirical formula mass** may be the same as the molar mass (RFM in grams and units are $g\,mol^{-1}$).

The formula that gives the correct molar mass is the molecular formula and it is the same as, or a multiple of, the empirical formula.

Determining formulae

When determining a formula you will be given either:

1 percentage compositions by mass of the elements (or water) in the compound (Assume you have 100 g and change the percentages to grams.)

2 mass data of the elements (or water) in the compound

3 mass measurements which allow you to calculate the mass of each element (or water) in the compound

Step 1: Work out the number of moles of each element (or water) in the compound.

Step 2: Divide each number of moles by the smallest number of moles (this sets the smallest number of moles to 1).

Step 3: If you have any fractions from Step 2, multiply all the numbers up to get whole numbers.

Step 4: Write the formula using the whole number mole ratio.

Worked example 1

A compound was found to have the following percentage composition by mass: carbon 35.0%, hydrogen 6.6% and bromine 58.4%. Determine its empirical formula.

Answer

$$\text{moles of carbon} = \frac{35.0}{12} = 2.917\,mol$$

$$\text{moles of hydrogen} = \frac{6.6}{1} = 6.6\,mol$$

$$\text{moles of bromine} = \frac{58.4}{80} = 0.73\,mol$$

→

The empirical formula shows the simplest whole number ratio of the atoms of each element in a compound.

The molecular formula shows the actual number of atoms of each element in one molecule of the substance.

Exam tip

Remember that the molecular formula is the same as, or a multiple of, the empirical formula.

Exam tip

Be careful with diatomic elements. A common mistake is to use the RFM of H_2 and Br_2. Only use the RAM as you want the number of moles of the atoms of each element.

The smallest number of moles is for bromine, so the simplest ratio is:

$$\text{bromine} = \frac{0.73}{0.73} = 1 \quad \text{hydrogen} = \frac{6.6}{0.73} = 9.04 = 9 \quad \text{carbon} = \frac{2.917}{0.73} = 3.996 = 4$$

Empirical formula is C_4H_9Br.

Worked example 2

A halogenoalkane was found to contain the following composition by mass: 35.96 g of carbon, 4.50 g of hydrogen and 159.54 g of chlorine. Determine the empirical formula of the halogenoalkane.

Answer

$$\text{moles of carbon} = \frac{35.96}{12} = 2.997\,\text{mol}$$

$$\text{moles of hydrogen} = \frac{4.50}{1} = 4.50\,\text{mol}$$

$$\text{moles of chlorine} = \frac{159.54}{35.5} = 4.494\,\text{mol}$$

The smallest number of moles is for carbon so the simplest ratio is:

$$\text{carbon} = \frac{2.997}{2.997} = 1 \quad \text{hydrogen} = \frac{4.50}{2.997} = 1.501 = 1.5 \quad \text{chlorine} = \frac{4.494}{2.997} = 1.499 = 1.5$$

Multiply by 2 to get a whole number ratio, i.e. 2:3:3.

Empirical formula is $C_2H_3Cl_3$.

Gas volume calculations

The **molar gas volume** (V_m) for any gas is the volume that 1 mol of the gas occupies.

At 20°C and 1 atm pressure $V_m = 24\,\text{dm}^3$ or $24\,000\,\text{cm}^3$.

The number of moles of a gas can be converted to gas volume by multiplying by the V_m:

gas volume = number of moles $\times V_m$

The gas volume can be converted to moles by dividing by the V_m:

$$\text{number of moles} = \frac{\text{gas volume}}{V_m}$$

Gas volume calculations are summarised in Figure 1.

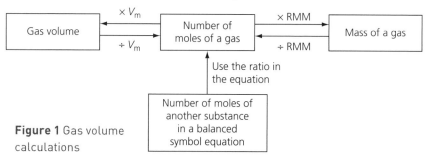

Figure 1 Gas volume calculations

Exam tip

The percentage values were given to one decimal place, so the ratios do not work out to exact whole numbers (e.g. 9.04 and 3.996). The closest whole number is used. Don't round too much. 1.499 is 1.5 rather than 1.

The **molar gas volume** is the volume of 1 mol of gas under specified conditions of temperature and pressure (for example, 24 dm³ at 20°C and 1 atm pressure)

Exam tip

The units of the gas volume should be the same as those of V_m: if you are using V_m in cm³, then the gas volume will be in cm³; if you are using V_m in dm³, then the gas volume will be in dm³.

Gas laws

Gay-Lussac's law states that in a reaction between gases, the volumes of the reacting gases, measured at the same temperature and pressure, are in a simple ratio to one another and to the volumes of any gaseous products. This law relates to the fact that when gases react, the volumes of gases in the reaction are in the same ratio as the balancing numbers in the equation for the reaction.

This law is derived from Avogadro's law, which states that 'equal volumes of gases under the same conditions of temperature and pressure contain the same number of particles'. For example:

$$CH_4(g) + 2O_2(g) \rightarrow CO_2(g) + 2H_2O(g)$$

This means that $10\,cm^3$ of methane react with $20\,cm^3$ of oxygen to give $10\,cm^3$ of carbon dioxide and $20\,cm^3$ of water vapour or $20\,cm^3$ react with $40\,cm^3$ to give $20\,cm^3$ and $40\,cm^3$.

This is true for all reactions involving gases, but only applies to the volumes of gases reacting or produced.

> **Knowledge check 1**
>
> What volume of oxygen reacts with $15\,cm^3$ of butane, C_4H_{10}, when it undergoes complete combustion?

Worked example

Calculate the volume of carbon dioxide, CO_2, produced when $25\,cm^3$ of propane gas, C_3H_8, are burnt completely in $200\,cm^3$ (an excess) of oxygen, O_2. Calculate the volume of oxygen remaining at the end of the experiment.

$$C_3H_8(g) + 5O_2(g) \rightarrow 3CO_2(g) + 4H_2O(g)$$

Answer

$25\,cm^3$ of propane require $125\,cm^3$ of oxygen to produce $75\,cm^3$ of carbon dioxide and $100\,cm^3$ of water vapour.

Therefore $125\,cm^3$ of the $200\,cm^3$ of O_2 are used. (Note that the propane volume is the limiting factor.)

So $(200 - 125 =)\ 75\,cm^3$ of the O_2 remain after the reaction.

Determining the formula of an unknown hydrocarbon

As can be seen from the worked example above, the combustion of gaseous hydrocarbons is a perfect application of this law. Experimental gas volume data from the combustion of an unknown gaseous hydrocarbon can be used to determine its formula.

Worked example 1

$20\,\text{cm}^3$ of an unknown gaseous hydrocarbon, C_xH_y, were mixed with $80\,\text{cm}^3$ (an excess) of oxygen in a graduated tube. A spark was applied and the resulting mixture of gases allowed to cool to room temperature. $60\,\text{cm}^3$ of gas remained. When this gas was exposed to sodium hydroxide solution, the volume reduced to $40\,\text{cm}^3$. Determine the formula of this hydrocarbon, C_xH_y.

Answer

Write a general equation for the combustion of a hydrocarbon. See Figure 2.

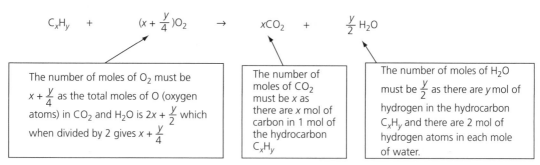

Figure 2 Equation for the combustion of a hydrocarbon

The initial mixture has a volume of $100\,\text{cm}^3$ and there is an excess of O_2, so all of the hydrocarbon, C_xH_y, is used and the mixture of gases after explosion contains O_2, CO_2 and $H_2O(g)$. The mixture is allowed to cool to room temperature, and the final $60\,\text{cm}^3$ consists of CO_2 and unused O_2 as the water vapour condenses.

When a gaseous mixture is exposed to sodium hydroxide solution (an alkali) the acidic gases in that mixture react with the sodium hydroxide and are removed into solution. CO_2 is the only acidic gas in the combustion mixture so the decrease in volume is due to its removal. The remaining volume after treatment with sodium hydroxide is the excess O_2.

Volume of $C_xH_y = 20\,\text{cm}^3$

Volume of $CO_2 = 20\,\text{cm}^3$ $(60 - 40)$

Volume of O_2 unused $= 40\,\text{cm}^3$

Volume of O_2 used $= 40\,\text{cm}^3$ $(80 - 40)$

Ratio of $C_xH_y : CO_2 = 20 : 20 = 1{:}1$, so $\boldsymbol{x = 1}$

Ratio of $C_xH_y : O_2$ used $= 20 : 40 = 1{:}2$,

so $= x + \dfrac{y}{4} = 2$ Therefore, $4x + y = 8$

$x = 1$, so $4 + y = 8$ and $\boldsymbol{y = 4}$

Formula of hydrocarbon is CH_4 (methane)

$100 \, cm^3$ of an unknown gaseous hydrocarbon, C_xH_y, were mixed with excess oxygen and a spark was applied. A contraction in volume of $200 \, cm^3$ was recorded after the mixture was cooled to room temperature. The mixture was treated with sodium hydroxide solution and a further contraction in volume of $200 \, cm^3$ was recorded. Determine the formula of the hydrocarbon C_xH_y.

The volume of CO_2 is $200 \, cm^3$ (reduction using sodium hydroxide solution).

Answer

In this example you are not given any volume of oxygen, so you must work out the volume of oxygen used by considering the gas volume changes during combustion (Figure 3).

| Change in volume during combustion (write as negative as it is a contraction) | = – | Volume of C_xH_y | – | Volume of O_2 used | + | Volume of CO_2 formed |

Figure 3 Changes in gas volume during combustion

$-200 = -100 -$ volume of O_2 used $+ 200$

Therefore, volume of O_2 used $= -200 + 100 - 200 = -300$, so volume of O_2 used $= 300 \, cm^3$

Volume of $C_xH_y = 100 \, cm^3$

Volume of $CO_2 = 200 \, cm^3$

Volume of O_2 used $= 300 \, cm^3$

Ratio of $C_xH_y : CO_2 = 100 : 200 = 1{:}2$ so $x = 2$

Ratio of $C_xH_y : O_2$ used $= 100 : 300 = 1{:}3$, so $x + \dfrac{y}{4} = 3$

Therefore, $4x + y = 12$

$x = 2$ and $8 + y = 12$, so $y = 4$

Formula of hydrocarbon is C_2H_4

Percentage yield

Not all chemical reactions go to completion.

The maximum mass of a product that could theoretically be obtained in a chemical reaction is called the **theoretical yield**.

The actual mass of the product obtained is called the **actual yield**.

The percentage yield is calculated as follows:

$$\text{percentage yield} = \frac{\text{actual yield}}{\text{theoretical yield}} \times 100$$

So

$$\text{theoretical yield} = \frac{\text{actual yield}}{\text{percentage yield}} \times 100$$

and

$$\text{actual yield} = \frac{\text{percentage yield} \times \text{theoretical yield}}{100}$$

Worked example 1

14.0 cm³ of butan-1-ol, C_4H_9OH (density 0.8 g cm⁻³) were used to prepare 1-bromobutane, C_4H_9Br. 9.5 g of 1-bromobutane were obtained. Determine the percentage yield. Give your answer to one decimal place.

Answer

The general method for this type of calculation is shown in Figure 4.

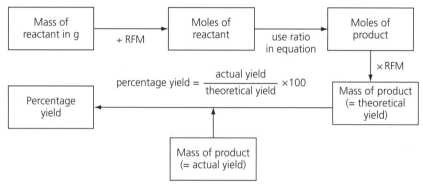

Figure 4 Calculation of percentage yield

Mass of butan-1-ol = $0.8 \times 14.0 = 11.2$ g

Moles of reactant = $\dfrac{11.2}{74} = 0.151$ mol (RMM of butan-1-ol = 74)

Moles of product = 0.151 mol (1:1 ratio in equation)

Theoretical yield of product = $0.151 \times 137 = 20.69$ g
(RMM of 1-bromobutane = 137)

Percentage yield $= \dfrac{9.5}{20.69} \times 100 = 45.9\%$

A question on percentage yield can be asked in reverse, where you have to determine the mass (or volume) of the reactant required to form a certain mass (or volume) of product based on a given percentage yield. This is more common at A2 but still can be asked at AS.

Worked example 2

12.7 g of magnesium nitrate were obtained from the reaction between magnesium and nitric acid, assuming a 77% yield. Calculate the mass of magnesium required, assuming an excess of nitric acid. ➡

Exam tip

The yields (actual and theoretical) may be mass values (which is more usual) or can be number of moles. Either is correct as long as they both have the same units.

Exam tip

This type of question may be structured with headings. If so, follow the headings carefully, putting appropriate units (including mol), as each will be worth a mark or may lose you a mark if you do not get it correct. Almost all reactions have a 1:1 ratio of reactant to product; if not, then this will be made clear in an equation. Because number of moles is calculated from mass and RFM, the volume of butan-1-ol needs to be converted to a mass value. Mass = density × volume.

Answer

$$Mg + 2HNO_3 \rightarrow Mg(NO_3)_2 + H_2$$

Actual yield of magnesium nitrate = 12.7 g

$$\text{Percentage yield} = \frac{\text{actual yield}}{\text{theoretical yield}} \times 100$$

$$\text{So theoretical yield} = \frac{\text{actual yield}}{\text{percentage yield}} \times 100$$

Theoretical yield of magnesium nitrate
(i.e. 100% yield) $= \dfrac{12.7}{77} \times 100 = 16.49$ g

Theoretical yield in moles of magnesium nitrate $= \dfrac{16.49}{148}$
= 0.111 mol
(RFM of magnesium nitrate = 148)

Moles of Mg required = 0.111 as 1:1 ratio of $Mg:Mg(NO_3)_2$

Mass of Mg required $= 0.111 \times 24 = 2.67$ g

Atom economy

Atom economy is a measure of how efficiently the atoms in the reactants are used and converted into the desired product in a chemical reaction.

Atom economy may be calculated as a percentage using the following expression:

$$\% \text{ atom economy} = \frac{\text{mass of desired product}}{\text{total mass of products}} \times 100$$

Worked example 1

The equation below shows the addition of bromine to cyclohexene.
Calculate the percentage atom economy for this reaction.

$$\begin{array}{ccccc} C_6H_{10} & + & Br_2 & \rightarrow & C_6H_{10}Br_2 \\ \text{cyclohexene} & & \text{bromine} & & \text{1,2-dibromocyclohexane} \end{array}$$

This is an addition reaction. Such reactions have a 100% atom economy because there is only one product.

Table 1

RFM of reactants		RFM of desired products		RFM of non-useful products	
C_6H_{10}	82	$C_6H_{10}Br_2$	242	—	—
Br_2	160	—	—	—	—
Total	242	Total	242	Total	0

$$\% \text{ atom economy} = \frac{242}{242} \times 100 = 100\%$$

Knowledge check 2

14.0 g of 1-bromopentane (RMM 151) were formed from 10.7 g of pentan-1-ol (RMM 88). Calculate the percentage yield to one decimal place.

Worked example 2

In a blast furnace, carbon monoxide is used to reduce iron(III) oxide to iron.

$$Fe_2O_3 + 3CO \rightarrow 2Fe + 3CO_2$$

iron(III) oxide   carbon monoxide   iron   carbon dioxide

Calculate the percentage atom economy for this reaction. Give your answer to one decimal place.

Table 2

RFM of reactants		RFM of desired products		RFM of non-useful products	
Fe_2O_3	160	2Fe	112	$3CO_2$	132
3CO	84	—	—	—	—
Total	244	Total	112	Total	132

$$\% \text{ atom economy} = \frac{112}{244} \times 100 = 45.9\%$$

Understanding atom economy

Chemists often use percentage yield to determine the efficiency of a chemical synthesis. A high percentage yield would indicate that the reaction is efficient in converting reactants into products. This is important for profit, but percentage yield does not take into account any waste products.

The chemical industry is concerned about its effect on the environment, particularly when it comes to waste. A reaction may have a high percentage yield, but have a low atom economy. This means that other products in the reaction are waste and with a high percentage yield there is a greater amount of waste.

In order to reduce waste, chemists are working towards the use of reactions that have both a high percentage yield and a high atom economy. This is the main thrust of green chemistry.

Summary

- The empirical formula shows the simplest whole number ratio of atoms of each element in the compound.
- The molar gas volume is $24\,dm^3$.
- Gas volume = number of moles × molar gas volume
- In the combustion of a hydrocarbon C_xH_y, the general equation is:

$$C_xH_y + (x + \frac{y}{4})O_2 \rightarrow xCO_2 + \frac{y}{2}H_2O$$

- The formula of an unknown gaseous hydrocarbon can be determined by measuring the volume of gases during combustion and after treatment with alkali.
- Percentage yield = $\dfrac{\text{actual yield}}{\text{theoretical yield}} \times 100$
- Atom economy = $\dfrac{\text{mass of desired product}}{\text{total mass of products}} \times 100$

■ Nomenclature and isomerism in organic compounds

Nomenclature is the system used to name organic compounds. The rules for nomenclature are based on the IUPAC (International Union of Pure and Applied Chemistry) system. The correct chemical name is often called the IUPAC name or the systematic name.

Types of formulae

There are **four** main types of formulae for organic molecules:

1   Molecular formula: in a molecular formula, the actual number of atoms of each element in the molecule is shown. For example:

- ethane is C_2H_6
- ethanol is C_2H_6O
- ethanoic acid is $C_2H_4O_2$
- propanone is C_3H_6O

2   Structural formula: in a structural formula, the structure of the molecule is shown, with all the atoms and covalent bonds (any ionic parts of an organic molecule may also be shown with their charges). Some examples are given in Figure 5.

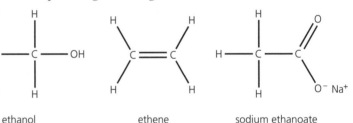

bromoethane   ethanol   ethene   sodium ethanoate

Figure 5 Examples of structural formulae

3   Condensed structural formula: this is written with the arrangement of atoms and groups at each carbon atom shown. Brackets are used to indicate that a group is bonded to a previous carbon atom and is not part of the main chain.

Example 1

Butanoic acid has the structural formula shown in Figure 6.

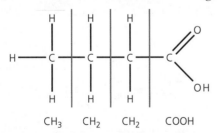

Figure 6 Structural formula of butanoic acid

The arrangement of atoms and groups for each carbon atom is determined. These are combined to give the condensed structural formula of butanoic acid: $CH_3CH_2CH_2COOH$.

Example 2

The structural formula of propan-2-ol is shown in Figure 7. Its condensed structural formula is: $CH_3CH(OH)CH_3$.

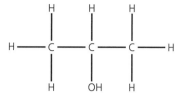

Figure 7 Structural formula of propan-2-ol

Example 3

The structural formula of 2,2-dichloropentane is shown in Figure 8. Its condensed structural formula is $CH_3CCl_2CH_2CH_2CH_3$.

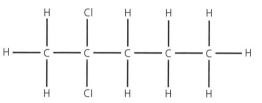

Figure 8 Structural formula of 2,2-dichloropentane

Example 4

The structural formula of 2,2-dimethylpentane is shown in Figure 9. Its condensed structural formula is $CH_3C(CH_3)_2CH_2CH_2CH_3$.

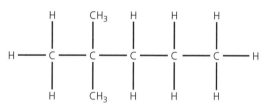

Figure 9 Structural formula of 2,2-dimethylpentane

4 Empirical formula: the empirical formula is the simplest ratio (cancelled down) of the atoms of the different elements in an organic compound. Some examples are shown in Table 3.

Table 3 Examples of empirical formulae

Compound	Molecular formula	Empirical formula
Ethanoic acid	$C_2H_4O_2$	CH_2O
Propanone	C_3H_6O	C_3H_6O
Butane	C_4H_{10}	C_2H_5

Note that the empirical formula and molecular formula of propanone are the same because it has only one oxygen atom, so the ratio cannot be cancelled down.

5 Skeletal formula: a skeletal formula shows the carbon backbone in a zigzag arrangement, omitting most hydrogen atoms, with only the end of a line

> **Exam tip**
>
> The OH group is shown in brackets as it is not part of the main chain and is bonded to a carbon atom in the middle of the chain. Individual atoms bonded to a carbon atom in the middle of the chain do not require a bracket.

> **Exam tip**
>
> There are three CH_3 groups on the left-hand side of this molecule, so another possible condensed structural formula could be $(CH_3)_3CCH_2CH_2CH_3$.
>
> Sometimes the C=C in alkenes and the C≡N in nitriles are shown in condensed structural formulae, e.g. CH_2=CH_2 instead of CH_2CH_2 for ethene and CH_3CH_2C≡N instead of CH_3CH_2CN for propanenitrile.

> **Exam tip**
>
> Empirical formulae are often determined from calculations. They are of limited use in writing equations, but you must be familiar with them.

representing the position of a carbon atom. Other functional groups are shown. Hydrogen atoms are only shown where necessary to avoid confusion with a carbon atom. Atoms that are not carbon or hydrogen are shown at the end of a bond. For OH and NH_2 groups, the OH and NH_2 are shown. Table 4 shows some examples of skeletal formulae.

Knowledge check 3

Give the systematic name for the following alkane:

$CH_3CH(CH_3)CH_2CH_2CH_3$

Table 4 Examples of skeletal formulae

Name	Structural formula	Skeletal formula
Ethane		
Propane		
Butane		
2-methylpentane		
1-bromobutane		
But-1-ene		
Z-but-2-ene		

Name	Structural formula	Skeletal formula
E-but-2-ene		
Buta-1,3-diene		
Ethanol		
Propan-1-ol		
Propan-2-ol		
Methanol		
Ethanal		
Propanal		

Name	Structural formula	Skeletal formula
Propanone		
Butanone		
Pentan-3-one		
Methanoic acid		
Ethanoic acid		
Methylamine		
Ethylamine		

Exam tip

Be careful with skeletal formulae; add hydrogen atoms where they are needed to avoid confusion with carbon atoms. Often C═C bonds are drawn horizontally, especially when showing E and Z isomers.

Naming alkanes, halogenoalkanes and alkyl-substituted alkanes

Alkanes have only single covalent bonds between their carbon atoms. The general formula for an alkane is C_nH_{2n+2}. Halogenoalkanes are alkanes in which one or more of the hydrogen atoms has been substituted by halogen atoms. Alkyl-substituted alkanes have one or more alkyl groups substituted in place of hydrogen atoms.

Step 1 Look for longest continuous carbon chain — this gives the main name (Table 5).

Exam tip

Substituent groups can appear on different molecules other than alkanes, but the rules for naming the substituent groups still apply.

Table 5 Naming alkanes

Number of carbon atoms	Alkane name	Formula	Structure of simplest straight-chain alkane
1	Methane	CH_4	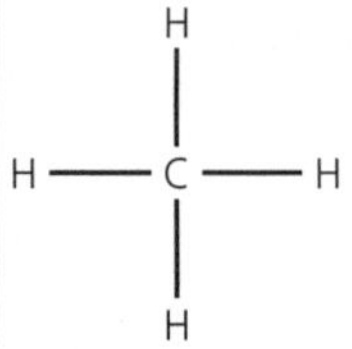
2	Ethane	C_2H_6	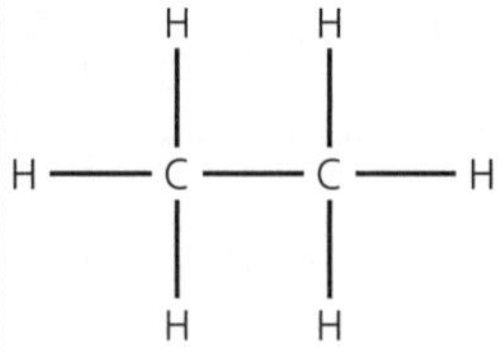
3	Propane	C_3H_8	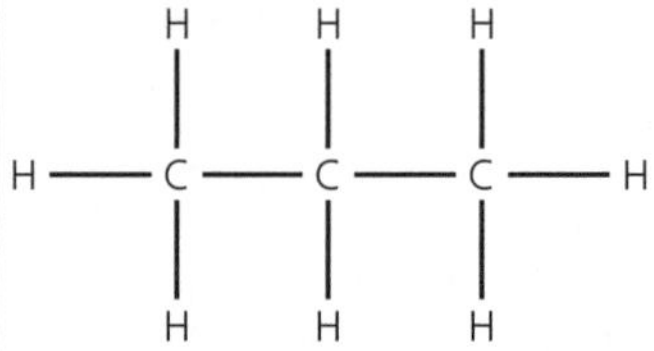
4	Butane	C_4H_{10}	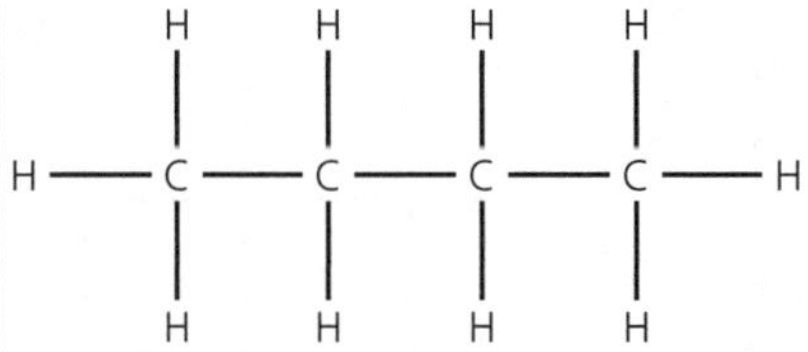
5	Pentane	C_5H_{12}	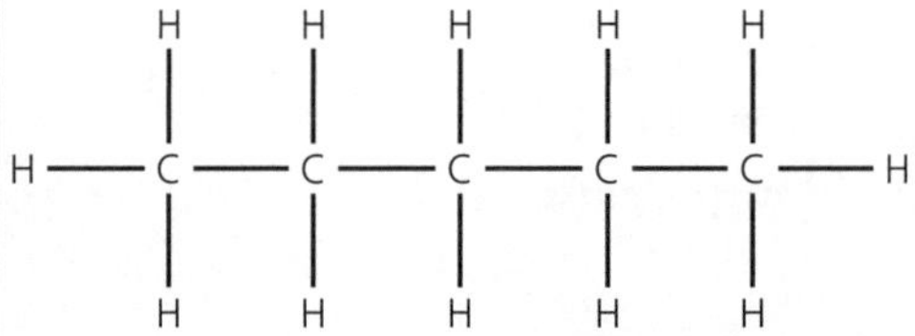
6	Hexane	C_6H_{14}	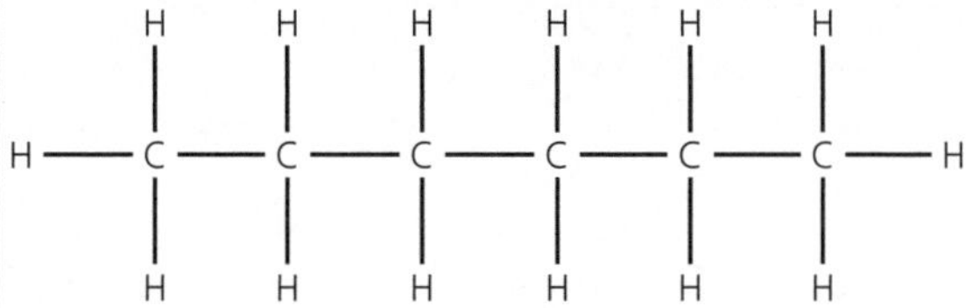

Exam tip

In exam questions, watch out for the longest chain not being drawn completely straight — this is common in questions and is there to catch you out. Always count the longest continuous chain.

Step 2 All substituent groups are named alphabetically before the main alkane name — for example, chloro before methyl. If there is more than one substituent group this is prefixed with di-, tri-, tetra- and so on — for example, dichloro if there are two chlorine atoms (even if they are bonded to different carbon atoms) and trimethyl if there are three methyl groups (again, even if each is bonded to a different carbon atom). Some common substituent groups are given in Table 6.

Step 3 Put a **locant number** in front of the substituent groups. A separate number is needed for every substituent group, with commas used to separate numbers: dichloro will require two locant numbers before it, one for the position of each chlorine atom: for example, 1,2-dichloro.

The carbon atoms in the longest chain are numbered from the end which will give the lowest locant numbers. Note that some substituents do not require a locant number if there is only one possible position for the substituent group(s) such as dichloromethane and bromoethane.

Step 4 Put dashes between letters and numbers.

Table 6 Common substituent groups

Group	Name
$-CH_3$	Methyl
$-C_2H_5$	Ethyl
$-C_3H_7$	Propyl
$-C_4H_9$	Butyl
$-F$	Fluoro
$-Cl$	Chloro
$-Br$	Bromo
$-I$	Iodo

Exam tip

The di-, tri- and tetra- prefixes do not change the alphabetical order — this is based on the name of the substituent group.

Example 1

- Figure 10 shows a halogenoalkane (an alkane with halogen atom(s) substituted in place of hydrogen).
- The longest continuous carbon chain is 2, so the name is derived from ethane.
- The substituent is bromo.
- A locant number is not needed, as there is only one position for a single bromo group.
- The name of this compound is **bromoethane**.

Figure 10

Example 2

- Figure 11 shows a halogenoalkane (an alkane with halogen atom(s) substituted in place of hydrogen).
- The longest continuous carbon chain is 3, so the name is derived from propane.
- The substituent is chloro.
- The locant number is 2, counting from either end.
- The name of this compound is **2-chloropropane**.

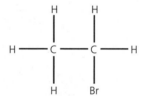

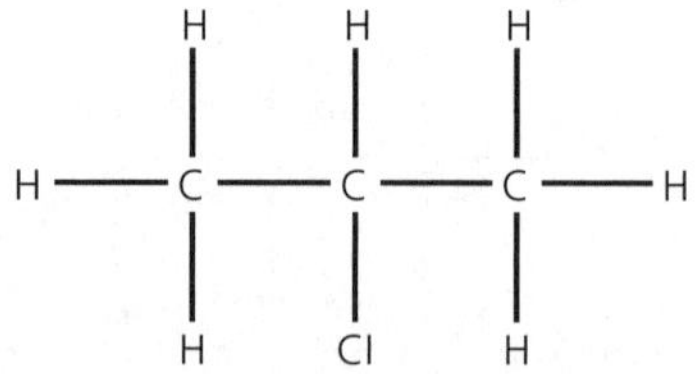

Figure 11

Example 3

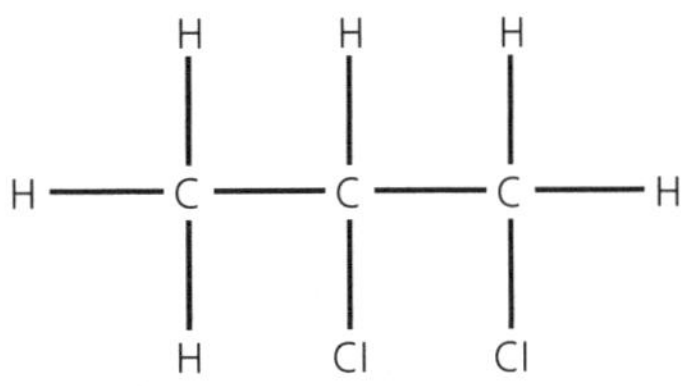

- Figure 12 shows a halogenoalkane (an alkane with halogen atom(s) substituted in place of hydrogen).
- The longest continuous carbon chain is 3, so the name is derived from propane.
- The substituent is chloro.
- There are two chloro substituent groups, so it is dichloro-.
- The locant numbers are 1 and 2, counting from the right-hand side.
- The name of this compound is **1,2-dichloropropane**.

Figure 12

Example 4

- Figure 13 shows an alkyl-substituted alkane (an alkane with alkyl group(s) substituted in place of hydrogen).
- The longest continuous carbon chain is 6, so the name is derived from hexane.
- The substituent is methyl.
- The locant number is 3, counting from the left-hand side.
- The name of this compound is **3-methylhexane**.

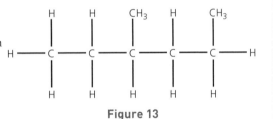

Figure 13

Exam tip

In this example watch out for the methyl group bonded to the right-hand carbon in the straight part of the chain — it is part of the longest continuous alkane chain.

Example 5

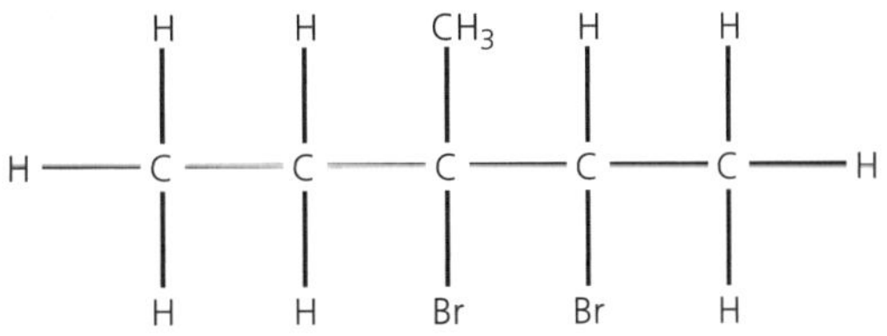

- Figure 14 shows an alkyl-substituted halogenoalkane.
- The longest continuous carbon chain is 5, so the name is derived from pentane.
- The substituents are bromo and methyl.
- There are two bromo substituent groups, so it is dibromo-.
- The locant numbers are 2,3,3, counting from the right-hand side.
- The name of this compound is **2,3-dibromo-3-methylpentane**.

Figure 14

Knowledge check 4

Give the systematic name for $CH_3CBr_2CH_2CH_3$.

Naming a molecule with a functional group

The common examples of molecules which contain a **functional group** at AS are given in Table 7 in order of decreasing nomenclature priority.

Table 7 Examples of molecules with a functional group

	Homologous series	Functional group	Nomenclature style and example
Highest priority	Carboxylic acid		**-oic acid** e.g. ethanoic acid
	Nitrile	—C≡N	**-nitrile** e.g. ethanenitrile
	Aldehyde		**-al** e.g. propanal
	Ketone		**-one** e.g. butanone
	Alcohol	——OH	**-ol** e.g. propan-1-ol
	Amine	——NH₂	**-amine** e.g. ethylamine
	Alkene		**-ene** e.g. propene
	Alkane		**-ane** e.g. propane
Lowest priority	Halogenoalkane	—C—X	Named as a substituted hydrocarbon e.g. chloroethane

A **functional group** is the part of a structure that determines the characteristic reactions of the compound.

> **Exam tip**
>
> Even though halogenoalkane contains the word 'alkane', the halogen atom of a halogenoalkane is a functional group because it determines the characteristic reactions of the compound.

> **Exam tip**
>
> The nomenclature of salts of carboxylic acid and sodium salts of alcohols will be dealt with in the individual sections relating to them.

> **Exam tip**
>
> The priority rule is important when a molecule has two or more functional groups as the group with the highest priority becomes the main name and the groups with the lower priority are named as substituent groups. Anything with lower priority than an alkane is named as a substituent group, such as chloro, bromo, iodo and methyl.

Rules for naming a molecule with a functional group are given below.

Step 1 Look for the longest carbon chain containing the functional group — this gives the main name.

Some **homologous series** require a locant number for the position of a single functional group. This applies to alkenes (with four or more carbon atoms), alcohols (with three or more carbon atoms) and ketones (with five or more

A **homologous series** is a series of compounds that have the same general formula and similar chemical properties, and successive members differ by CH_2.

carbon atoms). The presence of two identical functional groups may require a locant number for each, depending on the length of the carbon chain. The locant number for the highest priority functional group appears just before the ending, indicating the presence of the group, for example: pentan-2-one, but-1-ene, hexan-3-ol, ethane-1,2-diol, buta-1,3-diene.

Exam tip

When deciding whether or not to include a locant number, always think whether there is any other carbon atom in the molecule to which this substituent group or functional group could be bonded to make it a different molecule. If there is, then the name needs a locant number; if not, then no locant number is needed.

Step 2 Look for any other substituent groups and give them a locant number based on the end carbon atom closest to the highest priority functional group.

Step 3 Remember to include di-, tri-, tetra- if you have two, three or four respectively of the same substituent group.

Step 4 Put a **locant number** in front of the substituent groups. A separate number is needed for each substituent group, with commas used to separate numbers. Put dashes between letters and numbers.

Exam tip

The misplacing of commas and dashes will lose you marks, so make sure you separate numbers by commas and put a dash between numbers and letters.

Example 1

- Figure 15 shows a compound that contains an OH functional group, so it is an alcohol. Alcohols end in -ol.
- The longest continuous carbon chain containing the OH functional group is 3 and there are only single bonds between carbon atoms, so the name is derived from propane.
- A locant number is required for the OH group as it could be bonded to the second carbon atom.
- This compound is **propan-1-ol**.

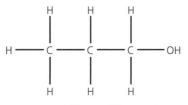

Figure 15

Example 2

- The compound shown in Figure 16 contains an OH functional group, so it is an alcohol. Alcohols end in -ol.
- The longest continuous carbon chain containing the OH functional group is 3 and there are only single bonds between carbon atoms, so the name is based on propane.
- A locant number is required for the OH group, as it could be bonded to the first carbon atom
- This compound is **propan-2-ol**.

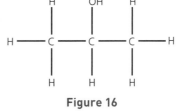

Figure 16

Exam tip

Propan-1-ol and propan-2-ol are structural isomers that differ in the position of the functional group. A locant number is required for a single OH group when the number of carbon atoms is three or more.

Example 3

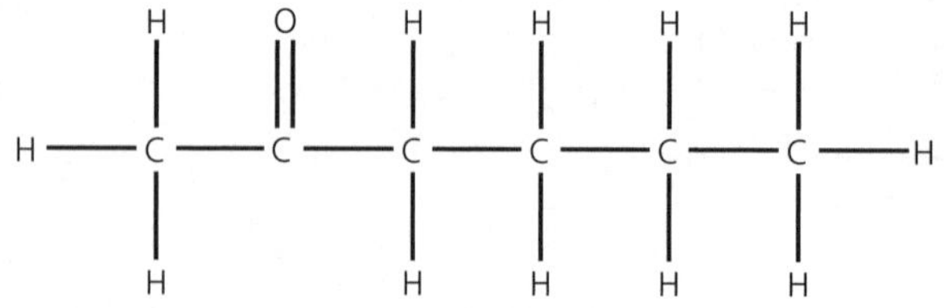

Figure 17

- Figure 17 shows a compound that contains a C=O functional group, so it is a ketone. Ketones end in -one.
- The longest continuous carbon chain containing the C=O functional group is 6 and there are only single bonds between carbon atoms so the name is based on hexane.
- A locant number is required for the C=O group as it could be bonded to the third carbon atom.
- This compound is **hexan-2-one**.

Example 4

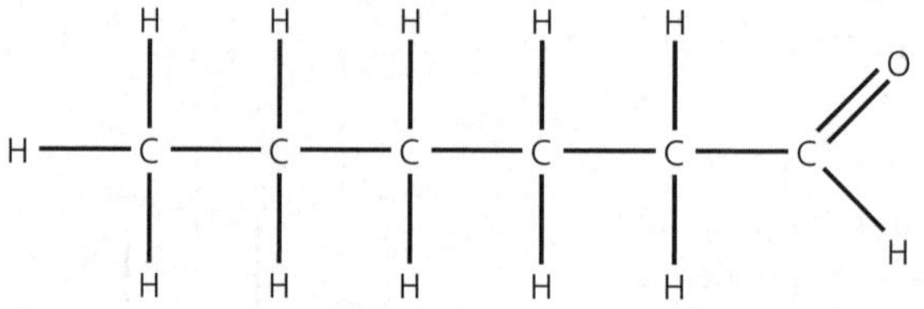

- Figure 18 shows a compound that contains a CHO functional group, so it is an aldehyde. Aldehydes end in -al.

Figure 18

- The longest continuous carbon chain containing the CHO functional group is 6 and there are only single bonds between these carbon atoms, so the name is based on hexane.
- A locant number is not required for the CHO group, as there is no other position for it.
- This compound is **hexanal**.

Example 5

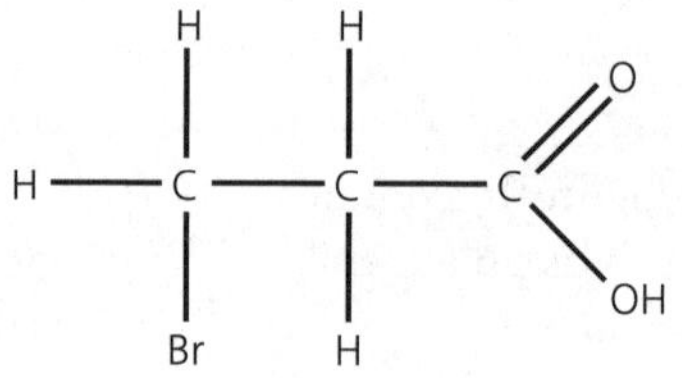

- In the compound in Figure 19, the highest priority functional group is COOH, so it is a carboxylic acid. Carboxylic acids end in -oic acid.
- The longest continuous carbon chain containing the COOH functional group is 3 and there are only single bonds between carbon atoms, so the name is based on propane.

Figure 19

- A locant number is not required for the COOH group as there is no other position for it.
- The chain is numbered from the highest priority group.
- The substituent group is bromo. Its locant number is 3, counting from the COOH group.
- This compound is **3-bromopropanoic acid**.

Exam tip

Hexan-2-one and hexanal (and hexan-3-one) are structural isomers that differ in their functional group. Hexan-2-one and hexanal are sometimes described as functional group isomers as they are from different homologous series but have the same molecular formula.

Example 6

- In Figure 20, the highest priority functional group is OH, so it is an alcohol. Alcohols end in -ol.
- The longest continuous carbon chain containing the OH functional group is 4 and there are only single bonds between carbon atoms, so the name is based on butane.
- A locant number is required for the OH group as it could be bonded to the first carbon.
- The chain is numbered from the end closest to the highest priority group.
- The substituent groups are dichloro (with locant numbers 4 and 4) and methyl (locant number 2).
- This compound is **4,4-dichloro-2-methylbutan-2-ol**.

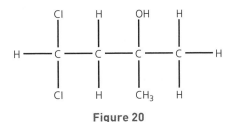

Figure 20

Exam tip

You must be able to:
- name molecules from structural formulae and from condensed structural formulae
- draw structural formulae from systematic names

Example 7

- In Figure 21, the highest priority functional group is C=C, so it is an alkene. Alkenes end in -ene.
- The longest continuous carbon chain containing the C=C functional group is 4 and, so the name is based on butene.
- A locant number is required for the C=C group as it could be between the second and third carbon atoms.
- The chain is numbered from the highest priority group.
- The substituent groups are dibromo (with locant numbers 3 and 4) and methyl (locant number 3).
- This compound is **3,4-dibromo-3-methylbut-1-ene**.

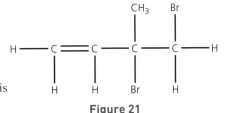

Figure 21

Multiple functional groups

If a molecule has two identical functional groups and there are other possible positions for the groups then they require locant numbers and di is added before the terminal name. If there are three identical functional groups, each may require a locant number and tri should be added before the terminal name.

Example 1

- The molecule shown in Figure 22 is an alkene.
- It has two alkene (C=C) groups.
- So the ending is -diene.
- The locant numbers for the C=C groups are 1 and 3.
- The molecule is called **buta-1,3-diene**.

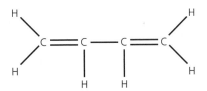

Figure 22

Exam tip

When a molecule has two (di) or three (tri) functional groups, a vowel (usually e or a) is added to the stem. Learn the examples given here as they are common.

Example 2

- The molecule shown in Figure 23 is an alcohol.
- It has two hydroxyl (OH) groups.
- So the ending is -diol.
- The locant numbers for the OH groups are 1 and 2.
- The molecule is called **ethane-1,2-diol**.

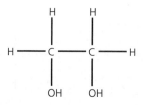

Figure 23

Example 3

- The molecule shown in Figure 24 is an alcohol.
- It has three hydroxyl (OH) groups.
- So the ending is -triol.
- The locant numbers for the OH groups are 1, 2 and 3.
- The molecule is called **propane-1,2,3-triol**.

Figure 24

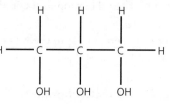

Knowledge check 5

Give the systematic name of the following molecule from its skeletal formula.

Structural isomerism

Structural isomerism is associated with organic compounds. Compounds that are structural isomers have the **same molecular formula** but they have **different structural formulae**. There are three ways in which this can occur:

- **Chain isomerism**, in which molecules have the same molecular formula but a different arrangement of the atoms in the carbon chain.
- **Positional isomerism**, in which molecules have the same functional group but it is in a different position in the molecule.
- **Functional group isomerism**, in which molecules have the same molecular formula but have different functional groups.

Structural isomers are molecules which have the same molecular formula but a different structural formula.

Chain isomerism

Chain isomerism occurs when there is more than one way of arranging the carbon skeleton of a molecule. For example, the carbon skeleton of an alkane of formula C_4H_{10} can be drawn as a straight chain or as a branched chain as shown in Figure 25.

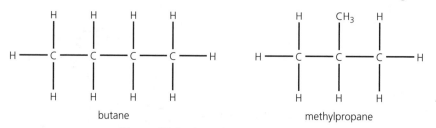

butane   methylpropane

Figure 25 Carbon skeletons of C_4H_{10}

Exam tip

Methylpropane is often called 2-methylpropane. There is no other position for the methyl group but you would not lose the mark for naming it if you included the 2.

C_5H_{12} exists as three different structural isomers, as shown in Figure 26.

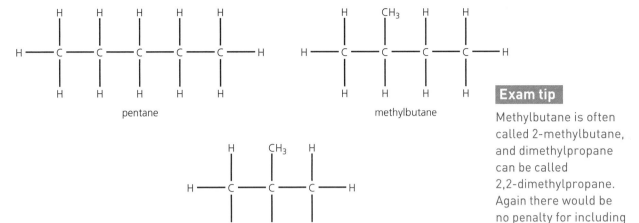

pentane

methylbutane

dimethylpropane

Figure 26 Structural isomers of C_5H_{12}

C_6H_{14} exists as five different structural isomers as shown in Figure 27.

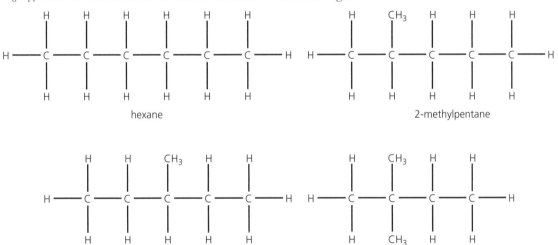

hexane

2-methylpentane

3-methylpentane

2,2-dimethylbutane

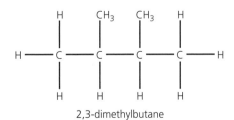

2,3-dimethylbutane

Figure 27 Structural isomers of C_6H_{14}

Positional isomerism

Positional isomers are isomers that differ in the position of the same functional group. Examples of positional isomers are given in Figures 28 and 29.

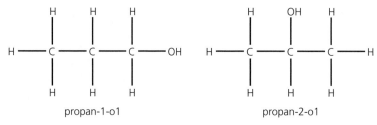

Figure 28 Positional isomers of propanol

The two molecules shown in Figure 28 (propan-1-ol and propan-2-ol) differ only in the position of the hydroxyl (OH) group. They are positional isomers.

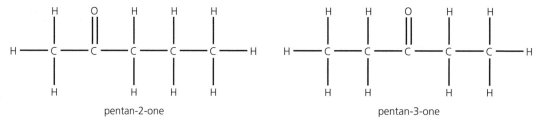

Figure 29 Positional isomers of pentanone

The two molecules shown in Figure 29 (pentan-2-one and pentan-3-one) differ only in the position of the carbonyl (C=O) group.

Functional group isomerism

Functional group isomers are a class of structural isomers that have the same molecular formula but have a different structural formula with a different functional group. Examples are shown in Figures 30 and 31.

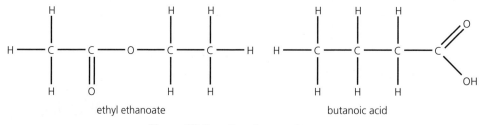

Figure 30 Functional group isomers

The molecules shown in Figure 30 have very different structures, but they have the same molecular formula ($C_4H_8O_2$). Ethyl ethanoate is an **ester** and butanoic acid is a **carboxylic acid**. Esters will be met during the A2 part of the course.

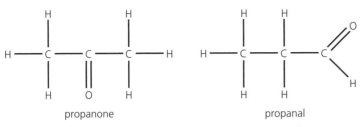

propanone   propanal

Figure 31 Propanone and propanal

The molecules shown in Figure 31 also have different structures, but they have the same molecular formula (C_3H_6O). Propanone is a **ketone** and propanal is an **aldehyde**.

Free rotation and geometric isomerism (stereoisomerism)

Stereoisomerism is the term used when two distinct isomers of a compound differ in the arrangement of the groups or atoms.

There are two types of stereoisomers — **geometric isomers** and optical isomers. Optical isomers are met at A2.

The C=C double bond in alkenes is an **energy barrier to free rotation** and it is the cause of geometric isomerism in alkenes. The C—C single bond in alkanes allows free rotation. The lack of free rotation about the double bond causes some alkenes and substituted alkenes to exhibit geometric isomerism. The presence of a C=C double bond does not imply that the molecule will exhibit geometric isomerism.

The geometric isomers in alkenes are referred to as *E* and *Z* isomers.

E–Z isomerism

- *Z* (from the German *zusammen*) means *together*.
- *E* (from the German *entgegen*) means *opposite*.

Whether a molecular configuration is designated *E* or *Z* is determined by the Cahn–Ingold–Prelog (CIP) priority rules (higher **atomic numbers** are given higher priority).

For each of the carbon atoms in the double bond, it is necessary to determine which of the two substituents is of higher priority.

If both of the substituents of higher priority are on the same side of the plane of the C=C bond, the arrangement is *Z*; if they are on opposite sides, the arrangement is *E*.

The CIP priority rules depend in the first instance on the atomic number of the atoms. Each atom bonded directly to the two carbons of the double bond is considered. They are ranked in order of decreasing atomic number (a priority list from highest to lowest). If two of the atoms are the same, the next atoms bonded to these same priority atoms are examined.

There are two classes of organic compounds that are functional group isomers of each other that you will come across:

- **Esters** are functional group isomers of **carboxylic acids**.
- **Aldehydes** are functional group isomers of **ketones**.

Geometric isomers are molecules which have the same structural formula but different arrangement of atoms due to the presence of one or more C=C bonds.

Exam tip

If the only difference in a group is an isotope (as they would have the same atomic number), mass numbers are used to determined priority.

Worked example 1

The molecules shown in Figure 32 are isomers of 1-bromo-2-chloro-1-fluoroethene.

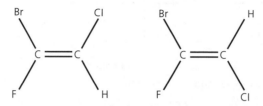

Figure 32 Isomers of 1-bromo-2-chloro-1-fluoroethene

Use the *E–Z* system by looking at the groups bonded to each carbon of the C=C bond (Figure 33).

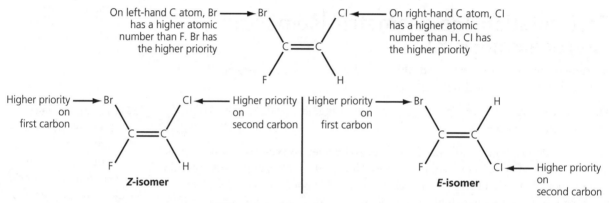

Figure 33 Using the *E–Z* system

Worked example 2

There are four isomers of C_4H_8 in total.

But-1-ene has a C=C double bond, but there are no geometric isomers of but-1-ene as it does not have two hydrogen atoms and two other groups bonded to each carbon of the C=C bond (Figure 34).

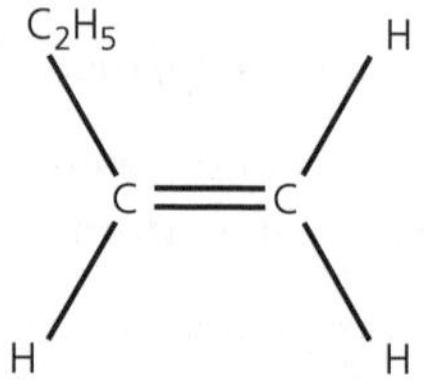

But-2-ene has a C=C double bond and there are *E–Z* isomers because it has two hydrogen atoms and two other groups bonded to each carbon.

Rotation of one end of the C=C double bond creates two different structures. These isomers are called *E* and *Z* isomers of but-2-ene (Figure 35).

Figure 34 But-1-ene

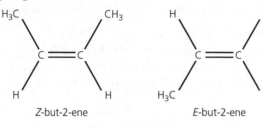

Z-but-2-ene   E-but-2-ene

Figure 35 But-2-ene

Exam tip

When looking at an alkene to decide if it would exist as *E–Z* isomers, look for two groups the same on one of the carbon atoms of the double bond — this means the molecule cannot exist as *E–Z* isomers. For example, $CH_3CH_2CH_2CH=CH_2$ cannot exist as *E–Z* isomers because the CH_2 at the right-hand side of C=C has two hydrogen atoms. This is a common multiple-choice question.

Methylpropene (or 2-methylpropene) (Figure 36) is a structural isomer of C_4H_8 which does not show E–Z isomerism as it does not have different groups bonded to each of the carbon atoms of C=C.

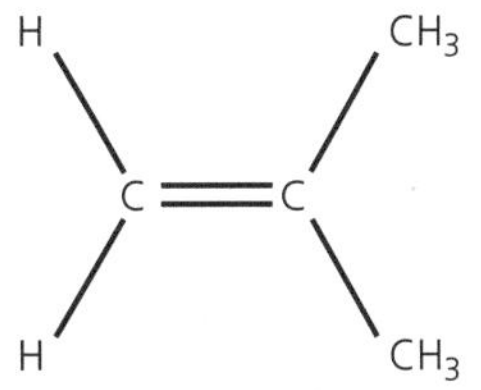

Figure 36 Methylpropene

Knowledge check 6

How many isomers of C_4H_8 exist?

Worked example 3

When the molecule has more than one single atom bonded, the naming system works in the same way, except that you look first at what atoms are bonded directly to each carbon atom in the double bond, then move one atom away from the double bond and look at what atoms are bonded there.

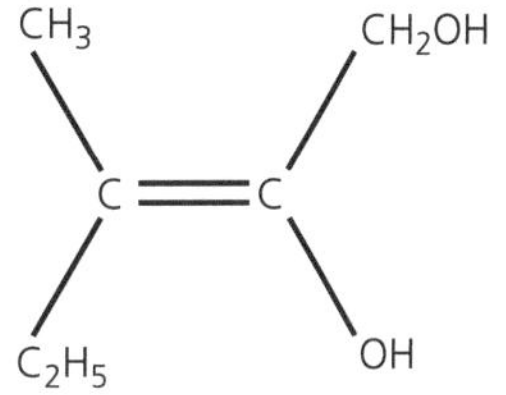

Figure 37

Consider the molecule shown in Figure 37.

Step 1 Look at the left-hand side first.

There are two carbon atoms bonded to the left-hand carbon atom of the C=C bond. These have the same atomic number, so they have the same priority. Now look at the atoms bonded directly to these carbon atoms. The top CH_3 group has three hydrogen atoms (total atomic number = 3) bonded to the carbon atom. The bottom C_2H_5 group has H, H and C (total atomic number = 8) bonded to the carbon atom. The C_2H_5 group a higher priority than the CH_3 group as at the second atom from the carbon of the C=C bond, there is a higher total atomic number.

Step 2 Look at the right-hand side.

The top CH_2OH group has C (atomic number 6) bonded directly to the right-hand carbon of the C=C bond, whereas the bottom OH group has O (atomic number 8) bonded to the carbon. The OH group has higher priority.

So the molecule shown in Figure 37 is the Z-isomer, as the two higher priority groups are on the same side of the plane of the C=C double bond.

Worked example 4

On the right-hand carbon atom of the C=C bond in the molecule shown in Figure 38, there is a group containing a double bond. The carbon atoms of the CHO and CH_2OH groups have the same priority. The atomic number of the atom at the end of the double bond must be counted twice. For example the CHO group has one hydrogen atom and one oxygen atom, but as there is a double bond between the carbon and the oxygen atoms, the oxygen atom must be counted twice, giving an atomic number

Figure 38

total of 17. CH_2OH has two hydrogen atoms and one oxygen atom, giving a total atomic number of 10. Therefore, CHO has higher priority than CH_2OH. C_2H_5 has higher priority on the left-hand carbon atom of the C=C bond, so the isomer is the Z-isomer.

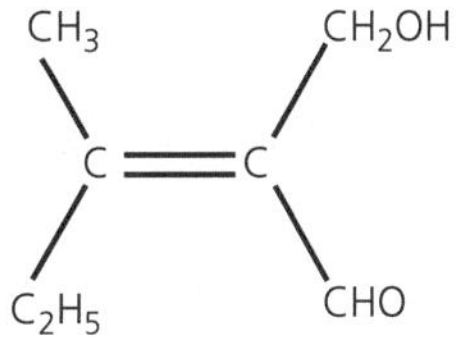

Cis–trans geometric isomerism

A simpler system for identifying these types of isomers is *cis–trans* isomerism. You should be familiar with it, as it is often used in organic and inorganic chemistry.

For *cis–trans* organic isomers to exist, there must be a hydrogen atom bonded to each of the carbon atoms of the C=C bond and another group bonded to each carbon atom which is not a hydrogen atom. Hydrogen atoms have the lowest priority, so this is a quick method of identifying the isomers. *Cis*-isomers are Z, *trans*-isomers are E. For example, see Figure 39. 1,2-dichloroethene (CHCl=CHCl) exists as *cis–trans* isomers. *Cis*-1,2-dichloroethene is the Z-isomer and *trans*-1,2-dichloroethene is the E-isomer.

Exam tip

E–Z is the system used for organic isomers, but it can also be applied to complex ions in A2. *Cis–trans* is also used to describe the C=C bond in fats.

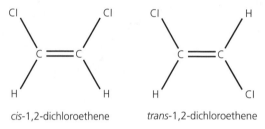

Figure 39 *Cis–trans* isomers of 1,2-dichloroethene

1,1-dichloroethene does not form *E–Z* (*cis-trans*) isomers.

Figure 40 summarises the different types of isomerism.

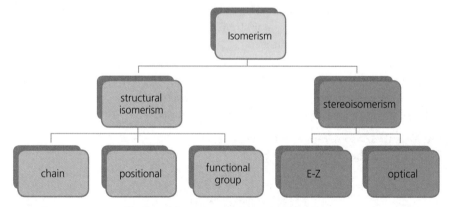

Figure 40 Different types of isomerism

Summary

- There are five main types of formulae used in organic chemistry — molecular, structural, condensed structural, skeletal and empirical.
- Nomenclature of alkanes, substituted alkanes and halogenoalkanes is based on the longest alkane continuous chain. Locant numbers are used to identify the position of any substituent groups (if necessary).
- Nomenclature of molecules from other homologous series is based on a terminal name ending, which depends on the functional group(s) present.
- Isomers are molecules with the same molecular formula but with a different structural formula or arrangement of atoms in space.
- The main types of isomerism are structural (chain, positional and functional group) and geometric (*E–Z* or *cis–trans*).

Alkanes

Alkanes and alkenes are the main **hydrocarbon** homologous series. Alkanes are described as **saturated** (they contain no C=C or C≡C bonds). The general formula for alkanes is C_nH_{2n+2}.

Boiling point of alkanes

Alkanes are non-polar and the only forces of attraction between neighbouring molecules are van der Waals forces. As the chain length of alkanes increases, the boiling point increases, as the van der Waals forces of attraction between the molecules increases. The physical state of the alkanes changes from gas (methane to butane) to liquid (pentane, C_5H_{12}, is the first liquid alkane) to solid. Heptadecane, $C_{17}H_{36}$, is the first alkane which is a solid at 20°C (melting point 21°C). The viscosity of the liquid alkanes also increases as the length of the carbon chain increases. The higher the branching in the molecule the lower the boiling point. More branching results in less van der Waals forces between the molecules because the molecules cannot pack as closely together. See Table 8.

Table 8 Boiling points of alkanes

Alkane	Molecular formula	Condensed structural formula	RMM	Boiling point/°C
Butane	C_4H_{10}	$CH_3CH_2CH_2CH_3$	58	−1
Methylpropane	C_4H_{10}	$CH_3CH(CH_3)CH_3$	58	−11.7
Pentane	C_5H_{12}	$CH_3CH_2CH_2CH_2CH_3$	72	36.1
Methylbutane	C_5H_{12}	$CH_3CH(CH_3)CH_2CH_3$	72	27.7

Combustion of alkanes

Combustion is the chemical reaction of fuels with oxygen, releasing heat and forming oxides. Combustion of fuels can be described as complete or incomplete. Complete combustion of fuels occurs in a plentiful supply of air and produces carbon dioxide and water. Incomplete combustion occurs when there is a limited supply of air and produces carbon monoxide, water and sometimes soot (unburnt carbon).

Worked example 1

Write an equation for the complete combustion of ethane.

Answer

$$C_2H_6 + 3\tfrac{1}{2}O_2 \rightarrow 2CO_2 + 3H_2O$$
or
$$2C_2H_6 + 7O_2 \rightarrow 4CO_2 + 6H_2O$$

Worked example 2

Decane ($C_{10}H_{22}$) burns in a limited supply of oxygen to produce carbon monoxide and water. Write an equation for this reaction.

Answer

$$C_{10}H_{22} + 10\tfrac{1}{2}O_2 \rightarrow 10CO + 11H_2O$$

A **hydrocarbon** contains only carbon and hydrogen atoms.

A **saturated hydrocarbon** contains no C=C or C≡C bonds.

Exam tip

You can work out the molecular formula for any alkane using the general formula. For example, an alkane with 17 carbon atoms, $C_{17}H_{2\times17+2}$, gives the molecular formula $C_{17}H_{36}$. You can also identify an alkane from the formula, because the number of hydrogen atoms should be twice the number of carbon atoms +2. This is helpful in identifying saturated molecules.

Exam tip

In worked example 1 either answer is acceptable. However, in energetics it is important to distinguish between these equations. If you are asked for an equation for the combustion of 1 mol of ethane, the first equation is the correct answer.

Combustion flames

Carbon burns with a smoky or sooty orange or yellow flame and hydrogen burns with a clean blue flame. Different organic compounds have different percentages of carbon by mass. Organic compounds with a high carbon content burn with a smoky orange flame. Organic compounds with a lower carbon content burn with a clean blue flame.

■ Alcohols with a low carbon number, such as ethanol (which has a percentage carbon content of 52.2%), burn with a clean blue flame (sometimes with an orange tip).

■ Alkanes and alkenes with a high carbon content (for example, hexane which has a percentage carbon content of 83.7%) burn with a smoky orange or yellow flame. This also shows that some **incomplete combustion** is occurring.

Pollution from fossil fuels

Burning fuels releases greenhouse gases (mainly carbon dioxide, CO_2) and acidic gases (sulfur dioxide, SO_2, and oxides of nitrogen, NO_x) into the atmosphere. SO_2 and NO_x come from combustion of impurities containing sulfur and nitrogen in the fossil fuels.

Greenhouse gases absorb infrared radiation emitted from the Earth's surface and reflect it back to the Earth. This increases the surface temperature of the Earth. The effects of increasing concentrations of greenhouse gases in the atmosphere are reported to be climate change, raised sea levels, melting of polar ice caps and flooding of low-level regions.

Carbon dioxide emissions have risen due to the increased combustion of fossil fuels. Carbon dioxide now makes up between 0.03 and 0.04% of the atmosphere.

Acidic gases dissolve into water in the atmosphere, resulting in acid rain. The effects of acid rain are corrosion of limestone features, statues and buildings, death of fish in lakes and rivers (due to pH changes) and defoliation of trees.

As pollution is a major environmental issue, many methods are used to limit the production of these greenhouse and acidic gases. These include:

■ removing sulfur from fuels before combustion
■ using catalytic converters on cars
■ scrubbing emissions (with alkali) from power stations and factories
■ using renewable energy sources — for example, wind, wave, solar, hydroelectric and tidal

Catalytic converters

The main polluting or harmful emissions from car exhausts are shown below, with the less pollution or harmful products to which they are converted. These conversions are achieved by either oxidation or reduction reactions on the surface of the catalytic converter.

- Carbon monoxide (CO) is oxidised to CO_2.
- Oxides of nitrogen (NO_x = NO or NO_2) are reduced to N_2 and O_2.
- Unburnt hydrocarbons (C_xH_y) are oxidised to CO_2 and H_2O.

The equations for these reactions are:

$$\left.\begin{array}{l} 2NO \rightarrow N_2 + O_2 \\ 2NO_2 \rightarrow N_2 + 2O_2 \end{array}\right\} \text{Reduction reactions}$$

$$\left.\begin{array}{l} 2CO + O_2 \rightarrow 2CO_2 \\ C_xH_y + (x + \frac{y}{4})O_2 \rightarrow xCO_2 + \frac{y}{2}O_2 \end{array}\right\} \text{Oxidation reactions}$$

Reactions of alkanes

Complete combustion

Conditions: a source of ignition and a plentiful supply of air (oxygen)

Products: $CO_2 + H_2O$

General equation: $C_xH_y + (x + \frac{y}{4})O_2 \rightarrow xCO_2 + \frac{y}{2}H_2O$

Example: $C_4H_{10} + 6\frac{1}{2}O_2 \rightarrow 4CO_2 + 5H_2O$

Incomplete combustion

Conditions: a source of ignition and a limited supply of air (oxygen)

Products: carbon (soot); $CO + H_2O$

Example: $C_2H_6 + 2\frac{1}{2}O_2 \rightarrow 2CO + 3H_2O$

Carbon monoxide is toxic and can be the cause of death by inhalation. Faulty domestic appliances that burn fossil fuels can produce carbon monoxide and so are dangerous.

Monohalogenation of alkanes

Alkanes react with halogens (chlorine and bromine) in the presence of ultraviolet light. The reactions produce a mixture of halogenoalkanes with varying numbers of halogen atoms. The reaction of methane with chlorine will produce a mixture of chloromethane (CH_3Cl), dichloromethane (CH_2Cl_2), trichloromethane ($CHCl_3$) and tetrachloromethane (CCl_4). The equations for these reactions are:

$$CH_4 + Cl_2 \rightarrow CH_3Cl + HCl$$

$$CH_3Cl + Cl_2 \rightarrow CH_2Cl_2 + HCl$$

$$CH_2Cl_2 + Cl_2 \rightarrow CHCl_3 + HCl$$

$$CHCl_3 + Cl_2 \rightarrow CCl_4 + HCl$$

Exam tip

Catalytic converters have a honeycomb ceramic structure coated in the metal catalyst. The metal coating keeps the cost down as less metal is used and the honeycomb structure gives it a larger surface area on which the reactions can take place. If leaded petrol were to be used in a car fitted with a catalytic converter the catalyst surface would become coated in lead which would render it passive (poison the catalyst) and stop it working.

Exam tip

For combustion equations of any organic molecules, you may be asked for complete combustion (forming CO_2 and H_2O) or incomplete combustion (usually forming CO and H_2O). It would be unusual to be asked for an equation forming soot (C) as one of the products of incomplete combustion, but you should be aware soot may be formed.

Mechanism: free radical photochemical substitution

The same reactions occur with bromine. The mechanism for the monohalogenation (first equation above) is most often studied. The mechanism (as detailed below) is the same for bromine as it is with chlorine, where Cl_2 and $Cl\bullet$ are replaced with Br_2 and $Br\bullet$, respectively.

The mechanism for the monohalogenation of alkanes is called free **radical photochemical** substitution. The term 'photochemical' indicates the need for ultraviolet light to initiate the reaction.

Initiation reactions:

For all initiation reactions:

    molecule $\rightarrow$ radicals

Here, the equation is:

$$Cl_2 \xrightarrow{\text{UV light}} 2Cl\bullet \quad (Cl\bullet = \text{chlorine radical})$$

This may be shown with the use of an arrow with half a head, showing the movement of one electron to each atom in **homolytic fission**:

$$\overset{\frown}{Cl}\overset{\frown}{}Cl \rightarrow 2Cl\bullet$$

This is an example of homolytic fission, as one molecule is split into identical species. That is, the bond is broken and one electron in the bond goes to each atom.

Propagation reactions:

For all propagation reactions:

    molecule + radical $\rightarrow$ radical + molecule

Here, the equations are:

$$CH_4 + Cl\bullet \rightarrow CH_3\bullet + HCl \quad (CH_3\bullet = \text{methyl radical})$$

and

$$CH_3\bullet + Cl_2 \rightarrow CH_3Cl + Cl\bullet$$

Termination reactions:

For all termination reactions:

    2 radicals $\rightarrow$ molecule

Here, the equations are:

$$Cl\bullet + Cl\bullet \rightarrow Cl_2$$

$$CH_3\bullet + Cl\bullet \rightarrow CH_3Cl$$

$$CH_3\bullet + CH_3\bullet \rightarrow C_2H_6$$

Exam tip

The mechanism for a reaction is the way that the reactants change into the products. Mechanisms may show the movement of electrons and the breaking and formation of covalent bonds, as well as any intermediate species.

Substitution is replacing one atom or group with a different atom or group.

A **radical** is a particle with an unpaired electron.

Homolytic fission is bond breaking in which one of the shared electrons goes to each atom.

Ethane (C_2H_6) is also one of the products of this reaction. All of the products are gaseous. Remember that that overall equation for the monohalogenation of methane is:

$$CH_4 + Cl_2 \rightarrow CH_3Cl + HCl$$

This mechanism can also explain the products of the reaction of any halogenoalkane with a halogen.

Worked example

Write an overall equation for the reaction of bromine with dibromomethane and write the initiation, propagation and termination steps in the mechanism.

Answer

Overall equation:

$$CH_2Br_2 + Br_2 \rightarrow CHBr_3 + HBr$$

The organic product is tribromomethane, $CHBr_3$.

Initiation reaction:

$$Br_2 \rightarrow 2Br\bullet$$

Propagation reactions:

$$CH_2Br_2 + Br\bullet \rightarrow CHBr_2\bullet + HBr$$

$$CHBr_2\bullet + Br_2 \rightarrow CHBr_3 + Br\bullet$$

Termination reactions:

$$2Br\bullet \rightarrow Br_2$$

$$CHBr_2\bullet + Br\bullet \rightarrow CHBr_3$$

$$2CHBr_2\bullet \rightarrow C_2H_2Br_4$$

$C_2H_2Br_4$ is 1,1,2,2-tetrabromoethane.

Knowledge check 7

Name all the gaseous products obtained when ethane undergoes monochlorination with chlorine in the presence of ultraviolet light.

Exam tip

It is important you can name 1,1,2, 2-tetrabromoethane as both of the radicals which form it have two bromine atoms, so when they bond, the substituted ethane has two bromo groups on each carbon atom. Think this through carefully.

Summary

- The general formula of the alkanes is C_nH_{2n+2}.
- The longer the chain of an alkane the higher the boiling point. The more branched an alkane the lower the boiling point.
- Alkanes undergo combustion and are used as fuels. Combustion of alkane fuels causes pollution.
- Carbon dioxide in the atmosphere is between 0.03 and 0.04%.
- The first four alkanes are gases, then liquid from C_5H_{12} up to $C_{16}H_{34}$. $C_{17}H_{36}$ is the first solid in the homologous series.
- Alkanes undergo combustion and react with halogens in the presence of ultraviolet light by a free radical photochemical substitution mechanism.

Alkenes

Alkenes are **unsaturated hydrocarbons** with the general formula C_nH_{2n}.

The C=C double bond in alkenes consists of a **sigma (σ)** covalent bond and a **pi (π)** covalent bond. All double bonds (C=O; O=O; C=C) consist of a **σ** covalent bond (often called a **σ-bond**) and a π covalent bond (often called a **π-bond**).

σ-bonds are stronger than π-bonds. The C=C bond is stronger than the C—C bond, but less than twice as strong.

The C=C **bond length** is shorter than the C—C bond length.

A σ-bond can be formed by linear overlap between two *s*-orbitals, resulting in free rotation along the bond axis.

Another possibility for a **σ**-bond is linear overlap between an *s*-orbital and a *p*-orbital, resulting in free rotation along the bond axis.

A σ-bond can also be formed by linear overlap between two *p*-orbitals.

A π-bond is formed by sideways overlap of *p*-orbitals. This results in restricted rotation along the bond axis.

The C=C bond in alkenes is a region of high electron density which makes it open to attack by electron-deficient species (electrophiles). This explains why alkenes are much more reactive than alkanes. When alkenes undergo **addition reactions** the π-bond is broken. It is said that addition occurs across the double bond.

> **Unsaturated hydrocarbons** contain at least one C=C or one C≡C bond.
>
> A **sigma (σ) bond** is a covalent bond formed by the linear overlap of atomic orbitals resulting in free rotation along the bond axis.
>
> **Bond length** is the distance between the nuclei of two covalently bonded atoms.
>
> A **pi (π) bond** is a covalent bond formed by the sideways overlap of *p*-orbitals, which restricts rotation along the bond axis.
>
> In an **addition reaction**, the π-bond of a double covalent bond is broken and two species add on across the double bond.

Straight-chain alkenes and their *E–Z* isomers

Table 9 Examples of straight-chain alkenes

n =	Name	Molecular formula	Structural formula	Physical state at RTP
2	Ethene	C_2H_4	H, H C=C H, H	Gas
3	Propene	C_3H_6	H—C=C—C—H H H H H *or* H, CH₃ C=C H, H	Gas

n =	Name	Molecular formula	Structural formula	Physical state at RTP
4	But-1-ene	C_4H_8	_or_	Gas
4	Z-but-2-ene	C_4H_8		Gas
4	E-but-2-ene	C_4H_8		Gas
5	Pent-1-ene	C_5H_{10}		Liquid
5	Z-pent-2-ene	C_5H_{10}		Liquid
5	E-pent-2-ene	C_5H_{10}		Liquid
6	Hex-1-ene	C_6H_{12}		Liquid

n =	Name	Molecular formula	Structural formula	Physical state at RTP
6	Z-hex-2-ene	C_6H_{12}		Liquid
6	E-hex-2-ene	C_6H_{12}		Liquid
6	Z-hex-3-ene	C_6H_{12}		Liquid
6	E-hex-3-ene	C_6H_{12}		Liquid

Dienes

A diene is an alkene with two C=C bonds in the main carbon chain.

- CH_2=C=CH_2 propadiene   (also known as allene)
- CH_2=C=CH–CH_3 buta-1,2-diene
- CH_2=CH—CH=CH_2 buta-1,3-diene
- CH_2=CH—CH_2–CH=CH_2 penta-1,4-diene

E–Z isomers may still exist at the different C=C groups. For example, hepta-1,3-diene, CH_2=CHCH=CHCH$_2$CH$_2$CH$_3$, can form E–Z isomers at the C=C bond positioned at the third carbon.

Exam tip

A symmetrical diene with the ability to form E–Z isomers at both C=C will only form three isomers, E at both C=C, Z at both C=C and the E–Z isomer will be the same as the Z–E isomer. Try drawing these out with 1,4-dibromobuta-1,3-diene.

Reactions of alkenes

Complete combustion

Conditions: a source of ignition and a plentiful supply of air (oxygen)

Products: $CO_2 + H_2O$

General equation: $C_xH_y + (x + \frac{y}{4})O_2 \rightarrow xCO_2 + \frac{y}{2}H_2O$

Example: $C_2H_4 + 3O_2 \rightarrow 2CO_2 + 2H_2O$

Incomplete combustion

Conditions: a source of ignition and a limited supply of air (oxygen)

Products: carbon (soot); $CO + H_2O$

Carbon monoxide is toxic and can cause death by inhalation of fumes from fires.

Example: $C_3H_6 + 3O_2 \rightarrow 3CO + 3H_2O$

Bromination of ethene (with HBr)

Product: C_2H_5Br (bromoethane)

Equation: $C_2H_4 + HBr \rightarrow C_2H_5Br$

Mechanism: electrophilic addition

> An **electrophile** is an ion or molecule that attacks regions of high electron density.

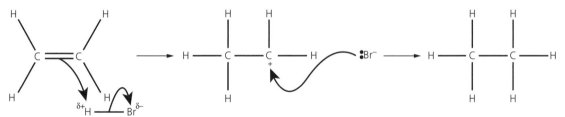

Figure 41 Electrophilic addition

Exam tip

The reaction of HBr with any C=C follows the same mechanism. The intermediate formed is called a carbocation. The reaction between the carbocation and the bromide ion is rapid as it is caused by ionic attraction. Make sure you can draw this mechanism in the three stages and show the polarity of the H–Br bond. The HBr undergoes heterolytic fission as the two species formed are differently charged ions.

The curly arrows in Figure 41 show the movement of a pair of electrons. The electrons move from the π-bond of the C=C to the hydrogen atom in HBr. The electrons in the H–Br bond move onto the bromine atom, leaving a bromide ion. This is an example of **heterolytic fission**.

The carbocation has a positive charge as it has lost an electron to the bromine atom. The lone pair of electrons on the bromide ion moves to the positive charge on the carbocation.

> **Heterolytic fission** is bond breaking in which the shared electrons go to one atom.

If the C=C is in a non-symmetrical larger molecule, there are two possible products when HBr is added. For example, HBr can react with but-1-ene to form 1-bromobutane or 2-bromobutane (Figure 42).

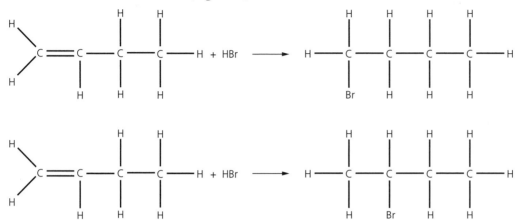

Figure 42 Reaction of but-1-ene with HBr

Both of these reactions work via the formation of a carbocation. The carbocations formed are shown in Figure 43. The carbocation on the left is a **primary carbocation** as there is one carbon atom bonded directly to the carbon atom with the charge; the carbocation on the right is a **secondary carbocation** as there are two carbon atoms bonded directly to the carbon atom with the charge.

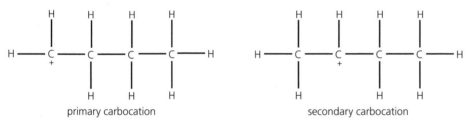

Figure 43 Carbocations formed in reaction of but-1-ene with HBr

Alkyl groups stabilise a carbocation. A tertiary carbocation is more stable than a secondary carbocation, which is more stable than a primary carbocation. The two alkyl groups stabilise the secondary carbocation so the major product is 2-bromobutane. Around 80% of the product is 2-bromobutane and 20% is 1-bromobutane.

A **primary carbocation** is a carbocation which has one carbon directly attached to the positively charged carbon.

A **secondary carbocation** is a carbocation which has two carbons directly attached to the positively charged carbon.

A **tertiary carbocation** is a carbocation which has three carbons directly attached to the positively charged carbon.

Knowledge check 8

Name the mechanism by which HBr reacts with ethene.

Worked example

3-methylpent-2-ene reacts with hydrogen bromide to form a major and a minor product. Draw and name the structures of the major and minor products and identify the major one. Explain why there is a major and a minor product.

Answer

The structure of 3-methylpent-2-ene is shown in Figure 44.

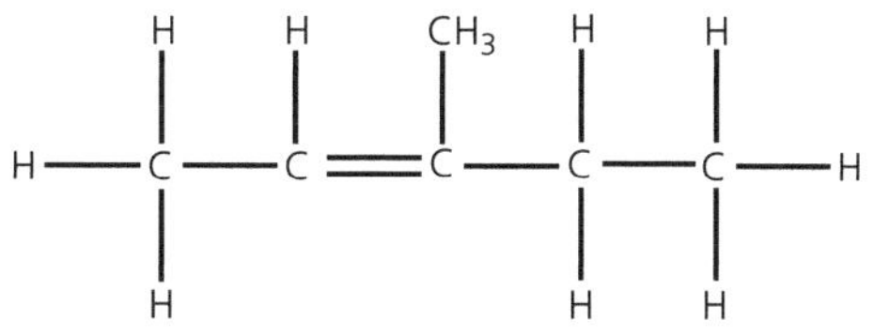

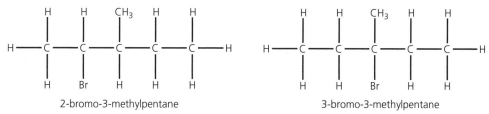

Figure 44 Structure of 3-methylpent-2-ene

When HBr reacts with 3-methylpent-2-ene there are two possible products which are shown in Figure 45.

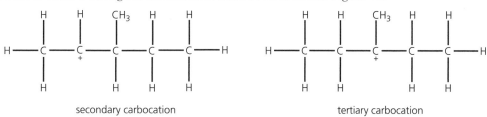

2-bromo-3-methylpentane   3-bromo-3-methylpentane

Figure 45 Products of reaction of HBr with 3-methylpent-2-ene

The carbocations formed during the mechanism would be as shown in Figure 46.

secondary carbocation   tertiary carbocation

Figure 46 Carbocations formed during reaction of HBr with 3-methylpent-2-ene

The major product is 3-bromo-3-methylpentane as the **tertiary carbocation** on the right is more stable than the secondary one on the left.

The chlorination of an alkene with HCl works by the same mechanism.

Hydrogenation of alkene (with H_2)

Conditions: finely divided nickel catalyst; 180°C; 4 atm

Product: an alkane

Equations:

$$C_2H_4 + H_2 \rightarrow C_2H_6$$
ethene   ethane

$$C_3H_6 + H_2 \rightarrow C_3H_8$$
propene   propane

Hydration of ethene (with steam)

Conditions: concentrated phosphoric acid (H_3PO_4); 300°C; 60–70 atm

Product: ethanol

Equation: $C_2H_4 + H_2O \rightarrow C_2H_5OH$
        ethene   ethanol

Hydrogenation is the addition of a hydrogen molecule across a C=C bond.

Exam tip

Hydrogenation is used to convert oils into solid fats. The C=C bonds in the oils are converted to C—C bonds. This process is called hardening or hydrogenation of oils.

Bromination with bromine (water), Br_2

Conditions: mix bromine water with a liquid alkene or bubble a gaseous alkene through bromine water

Product: a dibromoalkane

Figure 47

General equation (see Figure 47):

Example: the bromination of ethene to form 1,2-dibromoethane:

$$C_2H_4 + Br_2 \rightarrow C_2H_4Br_2$$

Structural formulae for this reaction are shown in Figure 48.

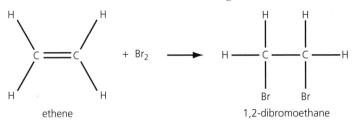

Figure 48

Test for a C=C bond

If yellow/orange/brown bromine water changes to colourless in the presence of an organic compound, the compound contains a C=C.

Addition polymerisation of alkenes

Ethene and other simple alkenes can undergo addition **polymerisation**. The polymerisation of ethene produces polythene (Figure 49).

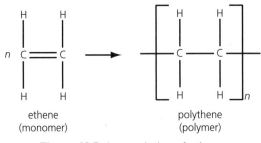

Figure 49 Polymersiation of ethene

> **Exam tip**
>
> The chlorination of an alkene with Cl_2 works in the same way. However, chlorine water is virtually colourless so this is not a valid test for unsaturation as it is difficult to detect any colour change. Iodine is unsuitable as it is less reactive than bromine and virtually insoluble in water.

Polymerisation is the joining together of small molecules (monomers) to form a large molecule.

Monomers are small molecules which join together to form a polymer.

A **polymer** is a large molecule formed when monomers join together.

Any addition polymerisation equation will be the same as above but the four groups attached to the carbon atoms may be different. For example, see Figure 50.

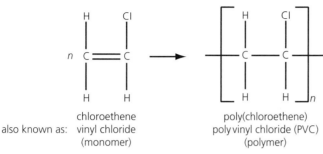

chloroethene
also known as: vinyl chloride
(monomer)

poly(chloroethene)
poly vinyl chloride (PVC)
(polymer)

Figure 50 An example of addition polymerisation

Use of 'excess reagent'

Sometimes questions are set on molecules with multiple functional groups — for example, two C=C or two OH groups. If an excess of a reagent is used, all of the functional groups will react in the usual way.

Figure 51 shows the reaction of penta-1,4-diene with an excess of bromine to give 1,2,4,5-tetrabromopentane. Both C=C bonds react when an excess of a reagent is present.

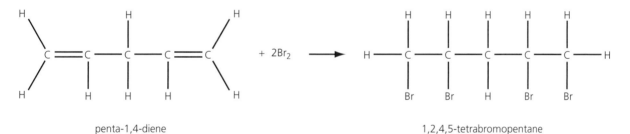

penta-1,4-diene

1,2,4,5-tetrabromopentane

Figure 51 Reaction of penta-1,4-diene with bromine

Summary

- The general formula of the alkenes is C_nH_{2n}.
- Alkenes contain a C=C functional group, which consists of a sigma covalent bond and a pi covalent bond.
- The double bond has a high electron density and is open to attack by electrophiles.
- Alkenes undergo combustion and also addition reactions due to the C=C bond.

- The addition reactions of alkenes occur by an electrophilic addition mechanism.
- Unsymmetrical alkenes such as but-1-ene react with HBr to form two different halogenoalkanes, one of which is the major product (2-bromobutane) and one the minor product (1-bromobutane).
- Alkenes can form addition polymers such as polythene.

Halogenoalkanes

The homologous series called the halogenoalkanes have the functional group C–X and the general formula is $C_nH_{2n+1}X$ where X is a halogen atom. (X = F, Cl, Br or I.)

Nomenclature of halogenoalkanes

Halogenoalkanes are named as substituted alkanes with the position of the halogen atom(s) indicated by locant numbers where necessary. Simple halogenoalkanes with up to two carbon atoms and one halogen atom do not require a locant number for the halogen atom. Simple halogenoalkanes with more than one carbon atom and two halogen atoms do require a locant number for each halogen atom. Halogenoalkanes derived from methane do not require locant numbers.

Primary, secondary and tertiary halogenoalkanes

Halogenoalkanes can be classified as one of the following based on the other atoms bonded to the carbon atom to which the halogen is bonded:

- primary (often written 1°)
- secondary (often written 2°)
- tertiary (often written 3°)

The simple way to work out the type of halogenoalkane is to count the carbon atoms bonded to the carbon atom that has the halogen bonded to it.

If the carbon atom bonded to the halogen atom has one carbon atom bonded to it, then the halogenoalkane is a primary halogenoalkane.

Examples of primary halogenoalkanes are shown in Figure 52.

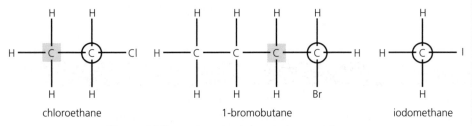

chloroethane   1-bromobutane   iodomethane

Figure 52 Examples of primary halogenoalkanes

The circled carbon atom is the carbon atom bonded to the halogen atom. In chloroethane and 1-bromobutane there is only one carbon atom (shown in a shaded box) bonded to the circled carbon atom, so they are primary halogenoalkanes. In iodomethane there are no carbon atoms bonded to the circled atom. It is also a primary halogenoalkane.

A **primary halogenoalkane** is a halogenoalkane that has a maximum of one carbon directly attached to the carbon which is bonded to the halogen.

A **secondary halogenoalkane** is a halogenoalkane that has two carbons directly attached to the carbon which is bonded to the halogen.

A **tertiary halogenoalkane** is a halogenoalkane that has three carbons directly attached to the carbon which is bonded to the halogen.

Exam tip

Halogenoalkanes with no alkyl groups bonded to the carbon atom which is bonded to the halogen atom are often classed as primary halogenoalkanes or can sometimes be referred to as methyl halogenoalkanes.

Knowledge check 10

Classify 2-chlorobutane as primary, secondary or tertiary.

If the carbon atom bonded to the halogen atom has two carbon atoms bonded to it then the halogenoalkane is a secondary halogenoalkane.

Examples of **secondary halogenoalkanes** are shown in Figure 53.

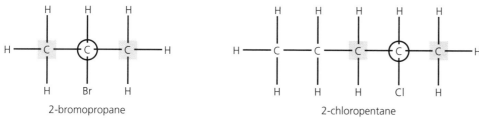

2-bromopropane   2-chloropentane

Figure 53 Examples of secondary halogenoalkanes

The circled carbon atom is the carbon atom bonded to the halogen atom. In both examples there are two carbon atoms (shown in shaded boxes) bonded to the circled carbon atom, so they are both secondary halogenoalkanes.

If the carbon atom bonded to the halogen atom has three carbon atoms bonded to it, then the halogenoalkane is a **tertiary halogenoalkane**. Examples are shown in Figure 54.

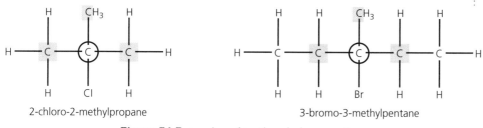

2-chloro-2-methylpropane   3-bromo-3-methylpentane

Figure 54 Examples of tertiary halogenoalkanes

The circled carbon atom is the carbon atom bonded to the halogen atom. In both examples there are three carbon atoms (shown in shaded boxes) bonded to the circled carbon atom, so they are both tertiary halogenoalkanes.

The classification of halogenoalkanes is important as the way they react depends on whether they are primary, secondary or tertiary.

Physical properties of halogenoalkanes

Electronegativity values of the halogen atoms are F = 4.0, Cl = 3.0, Br = 2.8 and I = 2.5. The strong polarity of the C–F and C–Cl bonds gives rise to permanent dipole–dipole attractions between chloroalkane molecules and between fluoroalkane molecules.

Van der Waals forces of attraction occur between all halogenoalkane molecules. If the molecule has symmetrical and equal polar bonds, e.g. CCl_4, it is non-polar and the only forces of attraction between molecules will be van der Waals forces.

> **Exam tip**
>
> Questions may require you to classify halogenoalkanes from molecular formulae, names, skeletal formulae or condensed structural formulae. You should quickly sketch the structural formulae of each compound and count the carbon atoms bonded to the carbon atom which is bonded to the halogen atom.

Trends in boiling point

Molecules of similar relative molecular mass (RMM) are often compared as they have a similar number of electrons and so similar van der Waals forces of attraction. Any significant difference in boiling point is then examined.

The boiling points of chloroalkanes (and fluoroalkanes) are slightly higher than those of alkanes of similar RMM (Table 10).

Table 10 Boling points of chloroalkanes and alkanes

Name	Formula	RMM	Boiling point/°C
1-chlorobutane	$CH_3CH_2CH_2CH_2Cl$	92.5	78.5
Hexane	$CH_3(CH_2)_4CH_3$	86	69

This is due to the permanent dipole–dipole attractions between chloroalkanes (and fluoroalkanes). They have similar van der Waals forces and the higher boiling point is due to breaking the permanent dipole–dipole attractions.

The boiling points of the halogenoalkanes derived from methane will increase as the mass of the halogen atom increases (Table 11).

Table 11 Boiling points of halogenoalkanes derived from methane

Name	Formula	RMM	Boiling point/°C
Chloromethane	CH_3Cl	50.5	–24
Bromomethane	CH_3Br	95	4
Iodomethane	CH_3I	142	42.5

The increase in boiling point is due to the increase in number of electrons in the halogen atom, which leads to greater van der Waals forces between the molecules.

As the carbon chain gets longer the boiling points of the halogenoalkanes increase (Table 12).

Table 12 Increase in boiling points of halogenoalkanes

Name	Formula	RMM	Boiling point/°C
Chloromethane	CH_3Cl	50.5	–24
Chloroethane	CH_3CH_2Cl	64.5	12.5
1-chloropropane	$CH_3CH_2CH_2Cl$	78.5	47
1-chlorobutane	$CH_3CH_2CH_2CH_2Cl$	92.5	79

The increase in boiling point with increasing carbon chain length is caused by the increase in van der Waals forces as there are more electrons and therefore more induced dipoles.

Exam tip

At room temperature and pressure, iodomethane is a liquid whereas chloromethane and bromomethane are gases. Chloroethane is a gas whereas bromoethane and iodoethane are liquids. All other simple halogenoalkanes are liquids.

The secondary halogenoalkanes (e.g. 2-chloropropane and 2-chlorobutane) have slightly lower boiling points than their 1-chloro isomers (Table 13).

Table 13 Boiling points of chloro isomers

Name	Formula	RMM	Boiling point/°C
1-chloropropane	$CH_3CH_2CH_2Cl$	78.5	47
2-chloropropane	$CH_3CHClCH_3$	78.5	36
1-chlorobutane	$CH_3CH_2CH_2CH_2Cl$	92.5	79
2-chlorobutane	$CH_3CH_2CHClCH_3$	92.5	68

This is because they cannot make as much contact with each other. The large chlorine atom in the middle of the molecule in the 2-chloro isomers reduces the contact between the hydrocarbon chains and this reduces van der Waals forces of attraction between the molecules.

Other physical properties

- Halogenoalkanes have a sweet, slightly sickly smell.
- Halogenoalkanes are immiscible with water and form two distinct layers. (Miscible liquids mix in all proportions, forming a single layer.)

Preparation of a halogenoalkane from an alcohol

Halogenoalkanes can be prepared by reacting an alcohol with a hydrogen halide, HX. HBr is used in the laboratory, but alcohols also react in the same way with HCl and HI.

For example, 1-bromobutane can be prepared from butan-1-ol:

$$CH_3CH_2CH_2CH_2OH + HBr \rightarrow CH_3CH_2CH_2CH_2Br + H_2O$$

butan-1-ol   1-bromobutane

and 2-bromopropane can be prepared from propan-2-ol:

$$CH_3CH(OH)CH_3 + HBr \rightarrow CH_3CHBrCH_3 + H_2O$$

propan-2-ol   2-bromopropane

The hydrogen bromide is generated in situ from the reaction of sodium bromide with concentrated sulfuric acid:

$$NaBr + H_2SO_4 \rightarrow NaHSO_4 + HBr$$

The overall equation can be written with NaBr and H_2SO_4:

$$CH_3CH_2CH_2CH_2OH + NaBr + H_2SO_4 \rightarrow CH_3CH_2CH_2CH_2Br + NaHSO_4 + H_2O$$

Method of preparation

The following is a standard method of preparation of a liquid halogenoalkane from the corresponding alcohol. The example is the preparation of 1-bromobutane from butan-1-ol (Figure 55).

Exam tip

From HF to HI the reactivity of the hydrogen halides towards alcohols increases. The reactivity of the alcohols increases from 1° to 2° to 3°.

Exam tip

'In situ' means that hydrogen bromide is not added directly to the reaction mixture. Other reagents are added which produce hydrogen bromide when they react.

Content Guidance

1 Mix sodium bromide with butan-1-ol and deionised water in a pear-shaped flask.

2 Put a reflux condenser in place above the flask and a still-head above the condenser. (*This maintains an open system and allows the concentrated sulfuric acid to drop smoothly into the flask.*)

3 Add concentrated sulfuric acid very slowly from a dropping funnel through the condenser and into the flask. (*This is a highly exothermic reaction and it is necessary to cool the contents of the flask by placing a beaker of cold water around it. Adding the concentrated sulfuric acid too quickly will result in dehydrating the alcohol, and some carbon (seen as black flakes) might appear in the flask.*)

4 Remove the dropping funnel and still-head and add anti-bumping granules to the flask. (*Anti-bumping granules promote smooth boiling.*)

5 Heat under **reflux** for 30 minutes (Figure 56). (*Organic reactions are slow because the covalent bonds are strong so the reaction needs to be heated. Most organic compounds are highly volatile and would boil off if heated (as the bonds between the molecules are weak). A reflux condenser is a condenser fitted vertically above a flask and there is continual evaporation and condensation without any loss of reactant or product. The condenser must be open at the top.*)

6 Distil off the product (Figure 57) until the upper organic layer has gone from the flask. (*Some water will distil over with the organic liquid and two layers will be seen in the measuring cylinder. The organic layer is mainly 1-bromobutane with some other impurities — water, unreacted butan-1-ol, sulfuric acid and hydrogen bromide.*)

Reflux is repeated boiling and condensing of a reaction mixture.

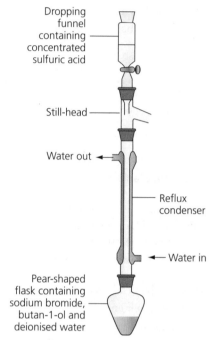

Figure 55 Preparation of 1-bromobutane

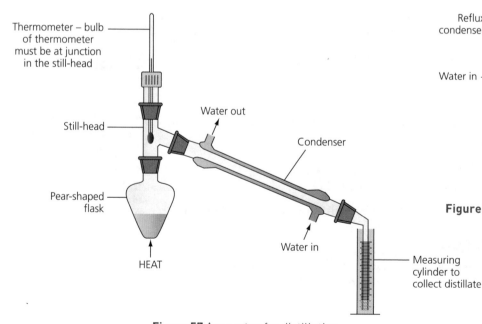

Figure 57 Apparatus for distillation

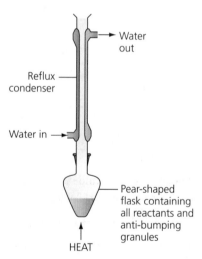

Figure 56 Apparatus for heating under reflux

7 Place the distillate in a clean separating funnel and separate the organic layer (*There are two ways to decide which is the organic layer:*

   a *Look up the density value of the halogenoalkane. If it is less dense than water (density of water = 1 g cm⁻³) it will be the top layer; if it is denser than water it is the bottom layer.*

   b *Add a small quantity of deionised water to the separating funnel. The layer that increases in volume is the aqueous layer.)*

8 Add concentrated hydrochloric acid to the impure organic product in a separating funnel. Stopper the separating funnel and shake to mix. Invert the separating funnel and open the tap while inverted to release any gas pressure. (*This process will remove the butan-1-ol impurity by converting the butan-1-ol to an ionic form, which is more soluble in the aqueous layer (the acid). The process of shaking an organic liquid with an aqueous solution is common in organic chemistry as a method of removing impurities from an organic liquid.)*

9 Clamp the separating funnel and allow the layers to separate (Figure 58). Remove the stopper and open the tap and allow the lower layer to run into a beaker. Run off the upper aqueous layer into another beaker and discard. (*The stopper must be removed before the tap is opened to run off the lower layer as it would be a closed system with the stopper in place and would prevent the layers from running out.)*

10 Put the lower organic layer into a clean separating funnel and wash with sodium hydrogencarbonate solution. Separate as before and discard the upper aqueous layer. (*The sodium hydrogencarbonate solution removes any acidic impurities in the organic liquid, such as hydrogen bromide and sulfuric acid. It is important to invert the separating funnel and release gas pressure by opening the tap, as carbon dioxide gas is formed in the reaction of the acids with sodium hydrogencarbonate.)*

11 The 1-bromobutane will be cloudy at this stage due to the presence of water. Add anhydrous sodium sulfate to the 1-bromobutane in small quantities, swirling after each addition until the 1-bromobutane is clear. (*Organic liquids will appear cloudy due to the presence of water and this is removed using a drying agent such as anhydrous sodium sulfate.)*

12 Decant or filter to remove the anhydrous sodium sulfate. (*Decanting is when the anhydrous sodium sulfate is allowed to settle to the bottom of the container and then the 1-bromobutane is poured off carefully into a clean beaker without disturbing the solid.)*

13 Distil the 1-bromobutane and collect between 101 and 103°C. (*The boiling point of 1-bromobutane is 102°C. When distilling as the final step in the purification of a liquid you should collect over a narrow range of temperature, usually 1°C maximum on either side of the boiling point. The range used to collect the product could be extended to collect more of the product as long as there are no other compounds with boiling points close to this range.)*

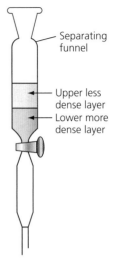

Separating funnel

Upper less dense layer
Lower more dense layer

Figure 58 Separating liquids of different densities

Exam tip

Many of the above points are asked in the practical exam at AS and A2. It is important to understand *why* practical procedures are carried out.

Reactions that produce halogenoalkanes

Halogenoalkanes can be prepared from alkanes, alkenes and alcohols.

From alkanes

General equation: alkane + halogen $\rightarrow$ halogenoalkane + hydrogen halide

Example: $CH_4 + Cl_2 \rightarrow CH_3Cl + HCl$

Conditions: ultraviolet light

Type of reaction: substitution

Mechanism: free radical photochemical substitution

From alkenes

General equation: alkene + hydrogen halide $\rightarrow$ halogenoalkane

Example: $C_2H_4 + HBr \rightarrow C_2H_5Br$

Conditions: none required — the gases react spontaneously

Type of reaction: addition

Mechanism: electrophilic addition

From alcohols

General equation: alcohol + HX $\rightarrow$ halogenoalkane + H_2O

Example: $CH_3CH_2CH_2CH_2OH + HBr \rightarrow CH_3CH_2CH_2CH_2Br + H_2O$

Conditions: HBr generated from NaBr + conc H_2SO_4 in situ; heat under reflux

Type of reaction: substitution

This is the standard method of manufacturing a halogenoalkane in the laboratory. Chloroalkanes can be manufactured from alcohols using PCl_5 (see page 59).

Reactions of halogenoalkanes

Reaction with alkali, OH⁻

General equation: halogenoalkane + OH^- $\rightarrow$ alcohol + X^-

Example: $CH_3CH_2CH_2Br + OH^- \rightarrow CH_3CH_2CH_2OH + Br^-$

      1-bromopropane   propan-1-ol

Conditions: heat under reflux with aqueous sodium hydroxide with a small quantity of ethanol for 1 hour

Type of reaction: substitution/hydrolysis

Mechanism: nucleophilic substitution

Exam tip

The reaction of alkanes with halogens is not a standard method of producing a halogenoalkane because there is no control over how many halogen atoms are substituted onto the alkane and there are other products (as seen from the mechanism on page 36).

Exam tip

The reaction of alkenes with hydrogen halides is not a standard method of producing a halogenoalkane because of the mixture of products obtained when longer chain alkenes are used, as the halogen atom may add at either end of the C=C bond.

In **hydrolysis**, molecules are broken up by reaction with water.

Hydrolysis

Hydrolysis reactions are often written as a reaction with water: For example:

$$C_2H_5Br + H_2O \rightarrow C_2H_5OH + HBr$$

Hydrolysis can be catalysed by either acid or base. Acid-catalysed hydrolysis usually involves heating the substance to be hydrolysed with a dilute mineral acid such as dilute hydrochloric acid. The reaction should be heated under reflux. The equation is written as the reaction with water with H^+ ions present over the arrow.

$$C_2H_5Br + H_2O \xrightarrow{H^+} C_2H_5OH + HBr$$

Base-catalysed hydrolysis involves heating the substance to be hydrolysed with dilute alkali. The equation may be written as the reaction with sodium hydroxide or hydroxide ions:

$$C_2H_5Br + NaOH \rightarrow C_2H_5OH + NaBr \quad or \quad C_2H_5Br + OH^- \rightarrow C_2H_5OH + Br^-$$

The mechanism of this reaction is described as nucleophilic substitution (often written as S_N).

The ease of hydrolysis of halogenoalkanes depends on the bond enthalpy of the C–X bond. The bond lengths and bond enthalpies of the C–X bonds are given in Table 14.

Table 14 Bond lengths and enthalpies of C–X bonds

Bond	Bond length/nm	Bond enthalpy/kJ mol^{-1}
C–Cl	0.177	327
C–Br	0.194	209
C–I	0.214	200

The longer the carbon–halogen bond, the lower the bond enthalpy. C–I bonds hydrolyse more readily than C–Br bonds, which hydrolyse more readily than C–Cl bonds.

This may be demonstrated experimentally. $2\,cm^3$ of ethanol is placed in three test tubes with $1\,cm^3$ of 1-chlorobutane in the first test tube, $1\,cm^3$ of 1-bromobutane in the second test tube and $1\,cm^3$ of 1-iodobutane in the third test tube. $1\,cm^3$ of silver nitrate solution is added to each test tube. The tubes are placed in a beaker of hot water.

First, a yellow precipitate (ppt) is observed in the third test tube, followed by a cream ppt in the second test tube and finally, and usually after a significant time, a white ppt in the first test tube.

Reaction with ammonia, NH$_3$

General equation: halogenoalkane + $NH_3 \rightarrow$ amine + HX (hydrogen halide)

Example: $C_2H_5Cl + NH_3 \rightarrow C_2H_5NH_2 + HCl$
          chloroethane   ethylamine

Conditions: heated with ethanolic ammonia in a sealed tube

Type of reaction: substitution

Mechanism: nucleophilic substitution

The NH_2 group substitutes for the halogen atom on the halogenoalkane.

> **Exam tip**
>
> In hydrolysis reactions, bonds in a molecule are broken by water molecules. One part of the molecule bonds with the H atom of water and the other part of the molecule bonds with the OH.

> **Exam tip**
>
> It is important to use halogenoalkanes with the same number of carbon atoms but different halogen atoms to ensure the test is fair. Different alkyl groups may have different effects on the carbon–halogen bond.

> **Exam tip**
>
> You may be asked to describe this procedure in terms of how to set it up and which precipitate forms first together with the colour of the precipitates. The ethanol is present as a common solvent for the halogenoalkanes and the aqueous solution of silver nitrate.

Reaction with cyanide ions, CN⁻

General equation: halogenoalkane + CN^- → nitrile + X^- (halide ion)

Example: C_2H_5Br + CN^- → C_2H_5CN + Br^-
bromoethane   propanenitrile

Conditions: halogenoalkane dissolved in ethanol dropped into an aqueous solution of sodium or potassium cyanide (KCN or NaCN). The mixture is heated under reflux.

Type of reaction: substitution

Mechanism: nucleophilic substitution

Reaction with potassium/sodium hydroxide in ethanol

General equation: halogenoalkane + KOH → alkene + KX + H_2O

Example: $CH_3CH_2CH_2Br$ + KOH → C_3H_6 + KBr + H_2O
1-bromopropane   propene

Conditions: halogenoalkane treated with hot ethanolic potassium (or sodium) hydroxide solution

Type of reaction: elimination

If the reaction of a halogenoalkane with either sodium hydroxide or potassium hydroxide dissolved in ethanol is asked for then the answer is of the type given above. However, a simpler answer would involve the elimination of the hydrogen halide as shown in Figure 59.

$$CH_3CH_2CH_2Br \xrightarrow[\text{heat}]{\text{KOH in ethanol}} C_3H_6 \ + \ HBr$$

Figure 59

Elimination from symmetrical and unsymmetrical halogenoalkanes

A symmetrical halogenoalkane is one in which the halogen atom is in the middle of the alkane molecule, such as 2-bromopropane. When the elimination reaction occurs with 2-bromopropane the product is propene.

When an unsymmetrical halogenoalkane such as 2-bromopentane is used, Figure 60 shows the possible products of the reaction.

Exam tip

The stars on the different coloured hydrogen atoms show all the hydrogens which may be eliminated with the neighbouring Br atom. The blue-starred H atoms lead to the formation of pent-2-ene and the red-starred H atoms lead to the formation of pent-1-ene. Pent-2-ene also forms *E–Z* isomers, so the reaction results in a mixture of pent-1-ene, *Z*-pent-2-ene and *E*-pent-2-ene.

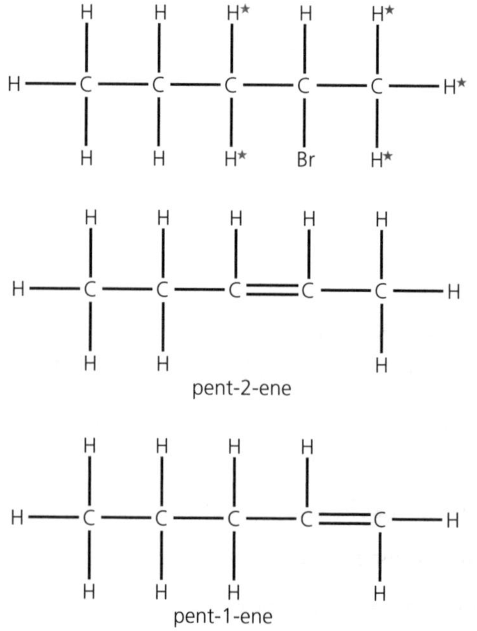

Exam tip

The reaction of a halogenoalkane with cyanide ions lengthens the carbon chain. Bromoethane is converted to propanenitrile. The nomenclature of amines and nitriles is studied at A2.

Exam tip

Ethanolic simply means dissolved in ethanol.

Elimination is a reaction in which a small molecule is removed from a larger molecule.

Write an equation for the reaction of 1-bromobutane with potassium hydroxide in ethanol.

pent-2-ene

pent-1-ene

Figure 60 Products of reaction of 2-bromopentane

Mechanisms

Four things are usually required when drawing the main mechanisms in AS chemistry:

1 initial reactants including the polarity of any relevant bond

2 an intermediate or a transition state

3 the products of the reaction

4 curly arrows to represent the movement of electrons

Nucleophilic substitution

For the two mechanisms discussed here, it is best to draw the halogenoalkane and alcohol tetrahedrally. The mechanism of the hydrolysis of halogenoalkanes using sodium hydroxide solution is described as nucleophilic substitution. Curly arrows should be used to show the movement of a pair of electrons. The attacking species is a nucleophile.

The mechanism of hydrolysis depends on the nature of the halogenoalkane.

Hydrolysis of primary halogenoalkanes

An example is the hydrolysis of bromoethane to form ethanol:

$$CH_3CH_2Br + OH^- \rightarrow CH_3CH_2OH + Br^-$$
bromoethane　　　　　ethanol

This mechanism is referred to as S_N2 because it is a bimolecular nucleophilic substitution (Figure 61). The S_N2 mechanism occurs for the hydrolysis of primary halogenoalkanes.

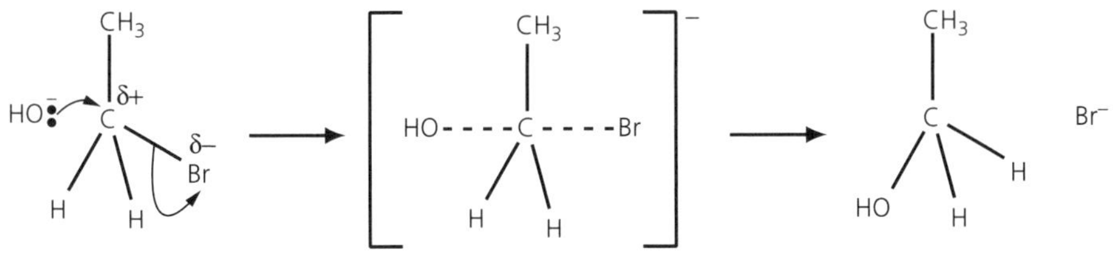

Stage 1

The attacking nucleophile OH⁻ transfers a lone pair of electrons from the oxygen atom to the δ+ carbon atom of the C–Br bond. This generates a transition state as shown in the next diagram.

Note that two species are involved in this first slow step of hydrolysis. This reaction is said to have a molecularity of 2 or to be **bimolecular**.

Stage 2

The transition state is formed where the molecule has two weak bonds with the attacking nucleophile (OH⁻) and the leaving group (Br).

There is a total of five bonds to the C atom so the transition state is highly unstable and cannot be isolated.

The reaction may reverse and reform the halogenoalkane and the attacking nucleophile.

Stage 3

The alcohol is formed and the bromide ion (Br⁻) remains in solution.

A group such as Br is often referred to as the **leaving group**.

Figure 61 Nucleophilic substitution of a primary halogenoalkane

A **nucleophile** is an ion or molecule, with a lone pair of electrons that attacks regions of low electron density.

Exam tip

Nucleophiles are often negatively charged ions such as OH⁻ or CN⁻ or a species with a lone pair of electrons such as NH_3. An electron-rich species is one that has more electrons than required in normal electron-pair bonding.

Hydrolysis of tertiary halogenoalkanes

An example is the hydrolysis of 2-bromo-2-methylpropane to form 2-methyl-propanol-2-ol:

$$(CH_3)_3CBr \quad + \quad OH^- \quad \rightarrow \quad (CH_3)_3COH \quad + \quad Br^-$$

2-bromo-2-methylpropane   2-methylpropanol-2-ol

This mechanism is referred to as S_N1 because it is a unimolecular nucleophilic substitution (Figure 62). The S_N1 mechanism occurs in the hydrolysis of tertiary halogenoalkanes.

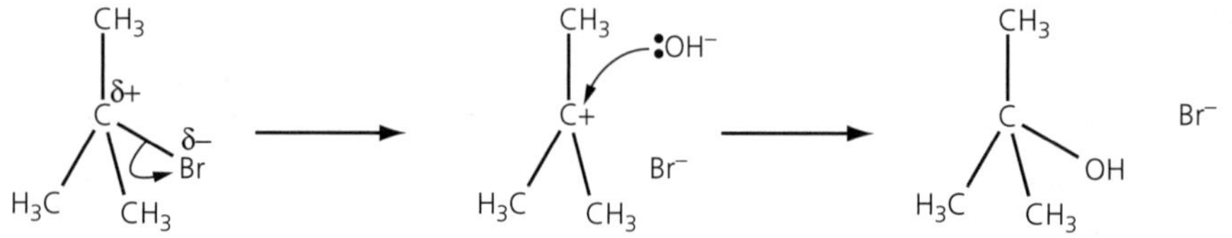

Stage 1

The tertiary halogenoalkane breaks down to form a carbocation and a bromide ion. Two electrons in the C–Br bond are transferred to the bromine to form a bromide ion. The carbocation is stabilised by the electron donating function of the three methyl groups attached to the main carbon atom.

Note that one molecule is involved in this first slow step of hydrolysis. This reaction is said to have a molecularity of 1 or to be **unimolecular**.

Stage 2

The carbocation (positively charged organic ion) is open to attack from nucleophiles.

The nucleophile (OH⁻) attacks the positively charged carbocation and produces the alcohol. The lone pair of electrons of the oxygen atom of the OH⁻ are donated to form the bond with the carbon atom.

Stage 3

The alcohol is formed and the bromide ion (Br⁻) remains in solution.

A group such as Br is often referred to as the **leaving group**.

Figure 62 Nucleophilic substitution of a tertiary halogenoalkane

Exam tip

The hydrolysis of secondary halogenoalkanes occurs via a combination of S_N1 and S_N2. You may have to recognise a nucleophile from a list of different species — look for the one with a lone pair of electrons.

Knowledge check 12

Name the mechanism by which 1-chloropropane would undergo hydrolysis.

Effect of CFCs on ozone

Ozone, O_3, is an allotrope of oxygen. Ozone absorbs harmful ultraviolet light, preventing a proportion of these harmful rays from reaching the surface of the Earth. Ultraviolet light that reaches the surface of the Earth can cause sunburn, but is also necessary for the production of vitamin D in humans.

In the absence of ozone, more ultraviolet light can reach the Earth. These wavelengths cause skin cancer, cataracts in eyes, damage to plant tissue and they reduce the plankton population in the oceans. Over the past three to four decades, scientists have observed a steady decrease in the ozone present in the stratosphere and have concluded that this decrease is caused by photochemical chain reactions by halogen free radicals.

The sources of these halogen free radicals are the halogenoalkanes, which have been extensively used as solvents, propellants, flame retardants and anaesthetics. Chlorine free radicals cause the greatest destruction due to the widespread use of chlorofluoroalkanes (CFCs), which were valued both for their lack of toxicity and non-flammability. Because CFCs are stable in the lower atmosphere they do not degrade. They diffuse into the upper atmosphere where ultraviolet light causes homolytic fission of the carbon–chlorine bond and produces a chlorine free radical, Cl•.

The chlorine radical reacts with O_3, forming a ClO• radical and oxygen gas:

$$Cl• + O_3 \rightarrow ClO• + O_2$$

Then the ClO• radical reacts with more ozone, forming more oxygen and another chlorine radical:

$$ClO• + O_3 \rightarrow 2O_2 + Cl•$$

The regenerated chlorine radical degrades more ozone. The use of CFCs is now restricted.

Summary

- The general formula for a halogenoalkane is $C_nH_{2n+1}X$.
- Halogenoalkanes can be classified as primary, secondary or tertiary, depending on the number of alkyl groups bonded directly to the carbon atom that has the halogen atom bonded to it.
- Halogenoalkanes can be prepared in the laboratory from alcohols, using hydrogen halides prepared in situ.
- Halogenoalkanes undergo substitution reactions in which the halogen atom is substituted for another group or atom — these occur by nucleophilic substitution.
- Halogenoalkanes undergo elimination reactions in the presence of ethanolic hydroxide ions.
- The products of the elimination reactions vary depending on the nature of the halogenoalkane.
- Primary halogenalkanes undergo hydrolysis by an S_N2 mechanism whereas tertiary halogenoalkanes undergo hydrolysis by an S_N1 mechanism.
- CFCs in the upper atmosphere destroy ozone, and this increases levels of ultraviolet radiation reaching the surface of the Earth.

▌Alcohols

Alcohols are saturated compounds containing the hydroxyl group (OH). The general formula of alcohols is $C_nH_{2n+1}OH$. The highly polar O–H bond allows alcohols to form hydrogen bonds with other alcohol molecules and with molecules such as water. Short-chain alcohols are soluble in water (these alcohols may also be described as being miscible with water). **Miscibility** is the ability of liquids to mix in all proportions. Strong hydrogen bonding between molecules causes a high boiling point, so small chain alcohols are liquids; higher ones are solid.

Miscibility is where liquids mix in all proportions. That is, they form a single layer.

Alcohols are classified in a similar way to halogenoalkanes as **primary**, **secondary** or **tertiary** (Figure 63):

A **primary alcohol** has one carbon directly attached to the same carbon as the OH group. (The exception to this is methanol, which is classed as primary.)

A **secondary alcohol** has two carbons directly attached to the same carbon as the OH group.

A **tertiary alcohol** has three carbons directly attached to the same carbon as the OH group.

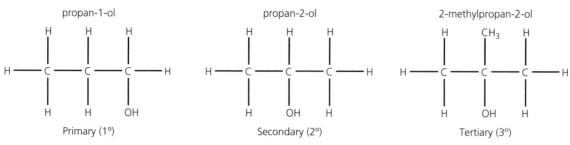

Figure 63 Primary, secondary and tertiary alcohols

Preparation of alcohols

- An alcohol can be prepared from the corresponding halogenoalkane by heating the halogenoalkanes under reflux with sodium hydroxide solution.
- This is a hydrolysis reaction and occurs by a nucleophilic substitution mechanism.
- A liquid alcohol can be distilled off and purified by fractional distillation.

Reactions of alcohols

The reactions of alcohols are varied. Some are reactions of the entire OH group; some are reactions of the H atom of the OH group and some involve complete disruption of the molecule.

Combustion of alcohols

Ethanol and other alcohols burn readily in a plentiful supply of air.

Equation: $C_2H_5OH + 3O_2 \rightarrow 2CO_2 + 3H_2O$

Conditions: heat and plentiful supply of oxygen

Products: $CO_2 + H_2O$

Observations: blue flame

In a limited supply of air, incomplete combustion of an alcohol can occur, producing carbon monoxide and/or carbon (soot) with water.

Alcohols as alternative fuels

- Alcohols burn readily, and ethanol may be used as an alternative to many hydrocarbon alkane fuels.
- 100 g of ethanol produces a possible 2979 kJ of energy with a possible 191 g of carbon dioxide, whereas 100 g of an alkane fuel such as propane (which has a similar RMM to ethanol) produces more energy (5004 kJ) but also more carbon dioxide (300 g).
- Carbon dioxide is a greenhouse gas which absorbs infrared radiation and re-emits it back towards the surface of the Earth. This causes global warming, changes in weather patterns, melting of the polar ice caps and flooding of low-lying land.
- Ethanol is considered to be a carbon-neutral fuel as the ethanol can be formed from fermentation of plant sugars made during photosynthesis. Photosynthesis uses up carbon dioxide in the atmosphere.

Exam tip

The halogenoalkane can also be mixed with 'moist silver oxide, which is a suspension of silver oxide, Ag_2O, in water. The mixture is boiled and the silver halide is precipitated out in the reaction. The alcohol can be distilled off and then purified by fractional distillation.

■ Also ethanol burns more cleanly with a clean blue flame while alkanes burn with a sooty yellow flame producing some soot which can contribute to the production of smog and also global dimming.

Reaction of alcohols with phosphorus pentachloride

Conditions: in a fume cupboard, add solid phosphorus pentachloride PCl_5 to the alcohol

Products: halogenoalkane; phosphorus oxychloride, $POCl_3$; hydrogen chloride, HCl (gas)

Observations: solid disappears; misty fumes; mixture warms up

For example, the production of chloroethane from ethanol and PCl_5 is shown in Figure 64.

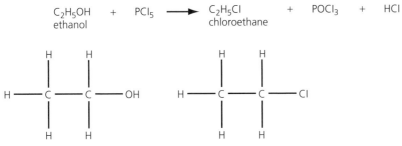

Figure 64 Production of chloroethane from ethanol and PCl_5

Reaction of alcohols with hydrogen bromide

Conditions: heat under reflux with hydrogen bromide (HBr). The HBr is formed in situ from sodium bromide and concentrated H_2SO_4 according to the equation:

$$NaBr + H_2SO_4 \rightarrow NaHSO_4 + HBr$$

Products: halogenoalkane + water

For example, the reaction of ethanol with HBr is shown in Figure 65.

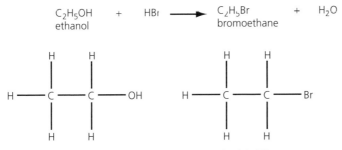

Figure 65 Reaction of ethanol with HBr

Reaction of alcohols with sodium

Conditions: a small piece of sodium metal added to an alcohol

Products: sodium alkoxide + hydrogen

Exam tip

PCl_5 reacts with any alcohol and converts the OH group to a Cl group. If used in **excess** with an organic compound containing more than one OH group, all the OH groups are converted to Cl. Watch out for the word excess in questions.

Knowledge check 13

Name all the products of the reaction of butan-2-ol with PCl_5.

Observations: fizzing

For example, the reaction of ethanol with sodium is shown in Figure 66.

$$2C_2H_5OH + 2Na \longrightarrow 2C_2H_5ONa + H_2$$

ethanol   sodium   sodium ethoxide   hydrogen

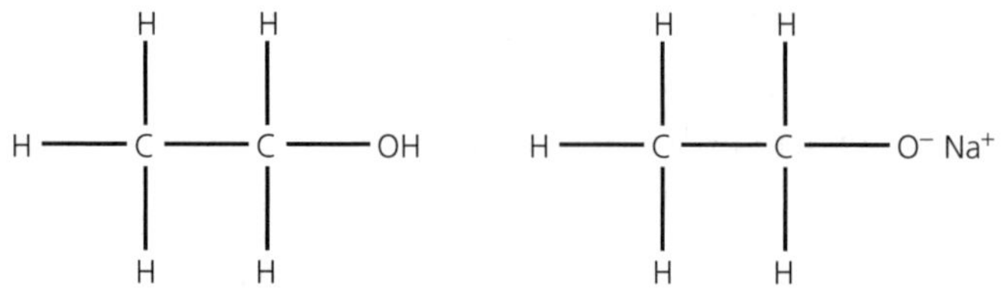

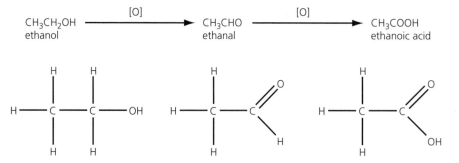

Figure 66 Reaction of ethanol with sodium

(Mild) oxidation with acidified potassium dichromate(VI)

- Acidified potassium dichromate(VI) solution (or acidified potassium dichromate solution) is a common oxidising agent used for 'mild oxidation' in organic chemistry.
- 'Mild oxidation' maintains the carbon chain length — there are no C–C bonds broken in mild oxidation.
- Sulfuric acid is used most commonly to acidify the solution as the sulfur in sulfuric acid is in its highest oxidation state (+6) and cannot be oxidised any further.
- Acidified potassium dichromate is often written as $H^+/Cr_2O_7^{2-}$, where H^+ represents the presence of the acid and $Cr_2O_7^{2-}$ represents the presence of the dichromate(VI) ion. The oxidation number of the chromium atoms in dichromate is +6.
- Primary, secondary and tertiary alcohols react differently when heated with acidified potassium dichromate solution.

(Mild) oxidation of a primary alcohol

A **primary alcohol** intially undergoes oxidation to an **aldehyde**. The aldehyde can be oxidised further to a **carboxylic acid**.

Conditions: heat with acidified potassium dichromate solution (heating under reflux yields the carboxylic acid whereas heating and distilling yields the aldehyde)

Observations: orange solution changes to green

An example is the oxidation of ethanol (Figure 67).

$$CH_3CH_2OH \xrightarrow{[O]} CH_3CHO \xrightarrow{[O]} CH_3COOH$$

ethanol   ethanal   ethanoic acid

Figure 67 Oxidation of ethanol

> **Exam tip**
>
> Use APDS to remember the initial letters of the (mild) oxidising agent: acidified potassium dichromate(VI) solution.

> **Exam tip**
>
> Often 'oxidation' is simply used to describe this process, omitting the term mild. Mild oxidation may be used to distinguish it from complete oxidation, which occurs in combustion and other reactions.

> **Exam tip**
>
> [O] represents an oxidising agent such as acidified potassium dichromate solution in organic reactions. [H] represents a reducing agent in organic reactions.

(Mild) oxidation of a secondary alcohol

Heating a **secondary alcohol** with acidified potassium dichromate solution produces a **ketone**. The orange solution changes to green.

An example is the oxidation of propan-2-ol (Figure 68).

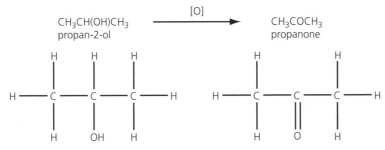

Figure 68 Oxidation of propan-2-ol

Tertiary alcohol

Tertiary alcohols are resistant to mild oxidation.

Writing oxidation reactions using [O]

[O] is used to represent the oxidising agent. If H atoms are removed, then the equation is balanced with H_2O on the right-hand side.

For example:

$CH_3CH_2OH + [O] \rightarrow CH_3CHO + H_2O$
    ethanol   ethanal

$CH_3CH_2OH + 2[O] \rightarrow CH_3COOH + H_2O$
    ethanol   ethanoic acid

$CH_3CH(OH)CH_3 + [O] \rightarrow CH_3COCH_3 + H_2O$
    propan-2-ol   propanone

> **Knowledge check 14**
>
> Write an equation using [O] to represent the oxidising agent for the mild oxidation of butan-2-ol to butanone.

Summary

- The general formula for an alcohol is $C_nH_{2n+1}OH$.
- Alcohols can be classified as primary, secondary or tertiary, depending on the number of alkyl groups bonded to the carbon to which the OH group is bonded.
- Alcohols undergo combustion and they may be used as an alternative to hydrocarbon alkane fuels.
- Alcohols react with phosphorus pentachloride and hydrogen halides, forming halogenoalkanes.
- Alcohols react with sodium, forming sodium alkoxide and hydrogen gas.
- Primary and secondary alcohols undergo (mild) oxidation using acidified potassium dichromate solution. Tertiary alcohols do not undergo (mild) oxidation.

■ AS organic identification tests

In practical examinations, tests can be carried out to identify and distinguish between organic chemicals. Table 15 summarises these tests.

Table 15 AS organic identification tests

Test and testing for	How to carry out the test	Typical observations	Deductions from observations
Appearance and smell	Observe colour and state; smell cautiously	Colourless liquid with characteristic spirit/alcohol smell	Possibly an alcohol/ethanol
		Colourless liquid with a sharp irritating smell	Possibly a carboxylic acid/ethanoic acid
Miscibility or solubility in water Testing for polarity of the liquid	Add a few cm^3 of the liquid to deionised water in a test tube	Mixes with water/one layer forms	Polar/can hydrogen bond with water — possibly an alcohol/carboxylic acid
		Does not mix with water/two layers form	Non-polar/cannot form hydrogen bonds with water
Combustion Testing for carbon content	Place a few drops of the liquid on a watch glass and ignite with a lit splint	Clean blue flame	Low carbon content/possibly an alcohol
		Sooty yellow/orange flame	High carbon content/possibly an alkane/alkene
Bromine water Testing for presence of C=C/alkene	Add a few cm^3 of bromine water to the liquid/solution in a test tube	Two layers form Bromine water changes from yellow/orange/brown to colourless	C=C/alkene present
		Two layers form Yellow/orange/brown colour of bromine water remains	No C=C present
Silver nitrate solution with ethanol Testing for presence of halogen atoms in a halogenoalkane	Add 1 cm^3 of the halogenoalkane to a few cm^3 of ethanol; add silver nitrate solution and warm in a water bath	White precipitate (slow to form)	Chloroalkane
		Cream precipitate (forms faster than white precipitate)	Bromoalkane
		Yellow precipitate (forms reasonably fast)	Iodoalkane
Phosphorus pentachloride (PCl_5) Testing for an OH/alcohol group	To a few cm^3 of the liquid add a few crystals of PCl_5 Test any gas released with a glass rod dipped in concentrated ammonia (NH_3) solution	Misty fumes; solid disappears; white smoke with concentrated ammonia solution	Alcohol/OH group present; HCl gas released; white smoke is ammonium chloride (NH_4Cl)
		No gas released; no white smoke with concentrated ammonia solution	Not an alcohol/no OH group present
Acidified potassium dichromate(VI) solution Testing for a primary/secondary alcohol as opposed to a tertiary alcohol	Mix a few cm^3 of the liquid or solution with acidified potassium dichromate solution in a test tube and warm in a water bath	Orange solution changes to green; change in smell	Primary or secondary alcohol group (OH) present
		Solution remains orange	Does not contain a primary or secondary alcohol group

Infrared spectroscopy

Infrared (IR) spectroscopy shows the bending and stretching (often combined to be called molecular vibrations) of covalent bonds. The bonds absorb energy in the infrared region of the electromagnetic spectrum. The wavelength of the radiation absorbed depends on the bond and on its environment. A molecular vibration in the lowest possible energy state is described as being in the ground state.

Features of an IR spectrum

The horizontal axis of an IR spectrum is wavenumber (measured in cm^{-1}) and ranging from about 400 up to 4000.

$$\text{wavenumber (cm}^{-1}) = \frac{1}{\text{wavelength}} \text{ (cm)}$$

The horizontal axis runs from 4000 cm^{-1} on the left to approximately 400 cm^{-1} on the right. The vertical axis is % transmittance. IR spectra are plotted as transmittance, so that there are dips in the plot from 100%.

The peaks and their range of wavenumbers shown in Table 16 are given in the Data Leaflet with your chemistry papers.

Table 16 Peaks and range of wavenumbers for IR spectra

Wavenumber/cm^{-1}	Bond	Compound
550–850	C—X (where X is Cl, Br, I)	Halogenoalkanes
750–1100	C—C	Alkanes, alkyl groups
1000–1300	C—O	Alcohols, esters, carboxylic acids
1450–1650	C=C	Arenes
1600–1700	C=C	Alkenes
1650–1800	C=O	Carboxylic acids, esters, aldehydes, ketones, amides, acyl chlorides
2200–2300	C≡N	Nitriles
2500–3200	O—H	Carboxylic acids
2750–2850	C—H	Aldehydes
2850–3000	C—H	Alkanes, alkyl groups, alkenes, arenes
3200–3600	O—H	Alcohols
3300–3500	N—H	Amines, amides

- The presence or absence of a peak is a clue to the identity of the compound — for example, a peak at between 1650 and 1800 cm^{-1} indicates the presence of a C=O bond in the molecule, suggesting an aldehyde, a ketone or a carboxylic acid (from A2 it may also be an ester, an acyl chloride or an amide).
- The absence of this peak would indicate the molecule does not contain a C=O bond and is therefore not one of the compounds listed above.
- A broad peak between 2500 and 3200 cm^{-1} indicates an O—H bond in a carboxylic acid — check for the presence of a C=O peak between 1650 and 1800 cm^{-1}.
- A narrower peak between 3200 and 3600 cm^{-1} indicates an O—H bond in an alcohol — check for the absence of a C=O peak between 1650 and 1800 cm^{-1}.

The **ground state** is the lowest possible energy state of a molecular vibration.

Knowledge check 15

What causes organic molecules to absorb infrared radiation?

Wavenumber is the frequency of a vibration measured in cm^{-1}

- The C–H peaks between approximately 2750 and 3000 cm^{-1} are present in almost all organic molecules.
- The region between 500 cm^{-1} and 1300 cm^{-1} is called the fingerprint region and is specific to a particular molecule. A molecule may be identified by comparing its IR spectrum to spectra in a database of known compounds.

An organic reaction can be monitored using IR spectroscopy. As the reaction proceeds, one functional group is changed into another so the signal for one group will decrease while the signal for another group will increase.

Exam tip

A simple analysis of an IR spectrum cannot identify a compound, but it can pinpoint functional groups which are present in the compound, leading to an identification of the homologous series to which the compound belongs. Further information or comparison with spectra from a database of known compounds could help identify the compound within the homologous series. Impurities in a sample can be detected using IR as the spectrum for the sample can be compared to the spectrum of the pure substance. The spectrum of an impure sample will have additional peaks.

A typical IR spectrum

The IR spectrum of ethanoic acid is shown in Figure 69.

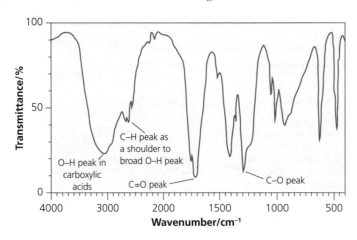

Figure 69 Example of an IR spectrum

Explanation

- The broad band between 2500 and 3200 cm^{-1} would suggest the presence of an O–H bond in a carboxylic acid.
- The peak at around 1700 cm^{-1} indicates the presence of a C=O bond, again adding to the evidence for a carboxylic acid.
- There is a C–O peak at around 1300 cm^{-1}.
- The C–H peak at around 2800 cm^{-1} is caused by C—H bonds.

Conclusion: The spectrum is for a carboxylic acid.

Summary

- Organic molecules absorb infrared (IR) radiation due to molecular vibrations (bending and stretching) of covalent bonds.
- An IR spectrum shows percentage transmittance against wavenumber (cm^{-1}).
- The absorption at various wavenumber values can shows the presence of a particular covalent bond within a molecule.
- The region between 500 and 1300 cm^{-1} is called the fingerprint region and can identify a molecule if compared to known spectra.

Energetics

Enthalpy is the total thermodynamic energy in a system. The system can be the reactants or the products in a chemical reaction.

ΔH is the **change in enthalpy** and the units of H and ΔH are kilojoules (kJ). For standard enthalpy changes, kJ mol^{-1} are used because the enthalpy change is quoted per mole of a particular substance in the reaction.

Chemical reactions are described as either exothermic or endothermic:

■ In an **endothermic reaction**, heat is taken in from the surroundings and ΔH is positive (enthalpy of the products is greater than the enthalpy of the reactants).

■ In an **exothermic reaction**, heat is given out to the surroundings and ΔH is negative (enthalpy of the products is less than the enthalpy of the reactants).

Simple enthalpy-level diagrams

Enthalpy-level diagrams show the relative enthalpy of the reactants and products, together with the change in enthalpy. The diagrams look different for endothermic and exothermic reactions. Because energy is taken in during an endothermic reaction, the products are at a higher enthalpy value, so the change in enthalpy (ΔH) is positive (+ve). Because energy is released during an exothermic reaction, the enthalpy decreases from reactants to products, so the change in enthalpy (ΔH) is negative (−ve) (Figure 70).

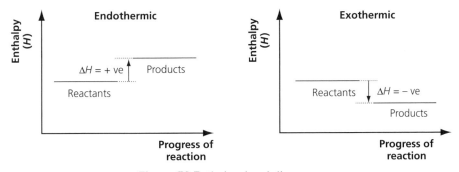

Figure 70 Enthalpy-level diagrams

Changes of state

Endothermic changes of state are melting, boiling, evaporation and sublimation as energy is taken in to cause the change in state.

Exothermic changes of state are condensation and freezing as energy is released during these changes of state.

An **endothermic reaction** is a reaction in which the enthalpy of the products is greater than the enthalpy of the reactants.

An **exothermic reaction** is a reaction in which the enthalpy of the products is less than the enthalpy of the reactants.

Exam tip

Make sure you can label an enthalpy-level diagram correctly — include the axes labels, reactants and products and show the correct increase or decrease in enthalpy. If it is for a specific reaction label the reactants and products lines with the appropriate reactants and products. The 'progress of reaction' axis may also be labelled 'reaction coordinate'.

Standard enthalpy changes

Standard enthalpy changes are the enthalpy changes for specific reactions that occur under **standard conditions**. Most enthalpy changes are measured in $kJ\,mol^{-1}$ (kilojoules per mole) and they are measured relative to 1 mol of a particular substance in the reaction. When an enthalpy change is under standard conditions, the symbol $\ominus$ is used as a superscript after ΔH.

> **Exam tip**
>
> This 'per mol' concept is very important for calculations because you can work out the energy change in a reaction based on standard enthalpy changes and the number of moles of the particular substance. The standard enthalpy change of combustion of ethanol is $-1367\,kJ\,mol^{-1}$. This means 1367 kJ of energy are released (the negative sign means the reaction is exothermic) when 1 mol of ethanol is burnt completely in oxygen. If 0.1 mol of ethanol were burnt completely in oxygen, 136.7 kJ of energy would be released.

When writing equations for reactions representing standard enthalpy changes, you must include state symbols for all reactants and products. All substances in the reaction are in their standard states at 298K and 100kPa.

Standard enthalpy of combustion

Symbol: $\Delta_c H^{\ominus}$ **Units:** $kJ\,mol^{-1}$

The equation representing the **standard enthalpy of combustion** of methane is:

$$CH_4(g) + 2O_2(g) \rightarrow CO_2(g) + 2H_2O(l)$$

1 mol of methane is burnt completely (forming carbon dioxide and water) and all substances are in their standard states at 298K and 100kPa pressure.

$\Delta_c H^{\ominus}$ (methane) $= -890\,kJ\,mol^{-1}$. This means that when 1 mol of methane is burnt completely in oxygen, 890 kJ of energy are released.

The equation representing the standard enthalpy of combustion of ethane is:

$$C_2H_6(g) + 3\tfrac{1}{2}O_2(g) \rightarrow 2CO_2(g) + 3H_2O(l)$$

1 mol of ethane is burnt completely (balanced for 1 mol of ethane); all substances are in their standard states. $\Delta_c H^{\ominus}$ (ethane) $= -1560\,kJ\,mol^{-1}$.

Standard enthalpy of neutralisation

Symbol: $\Delta_n H^{\ominus}$ **Units:** $kJ\,mol^{-1}$

The equation representing the **standard enthalpy of neutralisation** is:

$$H^+(aq) + OH^-(aq) \rightarrow H_2O(l)$$

1 mol of water is formed in a neutralisation reaction; all substances are in their standard states.

Standard conditions are 298K (25°C) and 100kPa.

A **standard enthalpy change** is the change in heat energy at constant pressure, measured at standard conditions.

> **Exam tip**
>
> Most texts and papers will refer to standard enthalpy changes without the term 'change'. For example, the standard enthalpy of combustion is taken as meaning the standard enthalpy change of combustion.

Knowledge check 16

What is meant by ΔH?

The **standard enthalpy of combustion** is the enthalpy change when 1 mol of a substance is completely burnt in oxygen under standard conditions.

> **Exam tip**
>
> Don't be afraid to use ½ and other fractions to balance an equation representing a standard enthalpy change. As the equation must be written for 1 mol of the substance to which it relates. In these combustion examples it has to be 1 mol of methane and 1 mol of ethane and the other substance are balanced based on that.

The standard enthalpy of neutralisation for a strong acid reacting with a strong alkali is $-57.6\,kJ\,mol^{-1}$. For a weak acid reacting with a strong alkali such as ethanoic acid with sodium hydroxide solution, the value is $-56.1\,kJ\,mol^{-1}$. The value is lower as the weak acid is not completely dissociated and some energy is required to complete this dissociation, so less energy is released as heat.

Standard enthalpy of formation

Symbol: $\Delta_f H^\ominus$ **Units:** $kJ\,mol^{-1}$

The equation representing the **standard enthalpy of formation** of ethanol is:

$$2C(s) + 3H_2(g) + \tfrac{1}{2}O_2(g) \rightarrow C_2H_5OH(l)$$

1 mol of ethanol is formed from its elements under standard conditions; all substances are in their standard states. $\Delta_f H^\ominus$ (ethanol) = $-277.7\,kJ\,mol^{-1}$.

The equation representing the standard enthalpy of formation of sodium nitrate is:

$$Na(s) + \tfrac{1}{2}N_2(g) + 1\tfrac{1}{2}O_2(g) \rightarrow NaNO_3(s)$$

1 mol of sodium nitrate is formed from its elements in their standard states; all substances are in their standard states. $\Delta_f H^\ominus$ ($NaNO_3$) = $-466.7\,kJ\,mol^{-1}$.

Standard enthalpy of reaction

Symbol: $\Delta_r H^\ominus$ **Units:** kJ

The **enthalpy of reaction** is the enthalpy change when substances react under standard conditions with the number of moles given by the equation for the reaction.

The equation representing the standard enthalpy of reaction for the hydrogenation of ethene is:

$$C_2H_4(g) + H_2(g) \rightarrow C_2H_6(g) \qquad \Delta_r H^\ominus = -138\,kJ$$

Experimental determination of enthalpy changes

Temperature change may be converted to energy change using the expression:

$$q = mc\Delta T$$

where q = change in energy in joules; m = mass in grams of the substance that undergoes the temperature change (usually water (for combustion) or solution (for neutralisation)) and c = specific heat capacity (energy required to raise the temperature of 1 g of a substance by 1°C) and ΔT = temperature change in °C or kelvin (K).

Determining enthalpy of neutralisation

Figure 71 shows the apparatus used to determine the enthalpy of neutralisation.

1   Add an exact volume of a known concentration of alkali into a polystyrene cup and measure the initial temperature.

2   Add an exact volume of a known concentration of acid to the polystyrene cup with stirring.

3   Measure the highest temperature reached and calculate the temperature change (ΔT).

The **standard enthalpy of neutralisation** is the enthalpy change when 1 mol of water is produced in a neutralisation reaction under standard conditions.

The **standard enthalpy of formation** is the enthalpy change when 1 mol of a compound is formed from its elements under standard conditions.

The **enthalpy of reaction** is the enthalpy change when the number of moles of substances are as written in the equation, under standard conditions.

Knowledge check 17

What are the standard conditions used in enthalpy changes?

Exam tip

For water, the specific heat capacity = $4.2\,J\,g^{-1}\,°C^{-1}$. For many solutions in neutralisation questions the value is assumed to be the same. Don't be concerned if the value is quoted as $4.2\,J\,g^{-1}\,K^{-1}$ as a temperature change of 1 kelvin (K) is the same as a temperature change of 1°C.

4 Calculate the energy change using $q = mc\Delta T$.

5 Calculate the energy change for 1 mol of water by dividing by the number of moles of water produced in the neutralisation reaction and then divide by 1000 to obtain a value in $kJ\,mol^{-1}$.

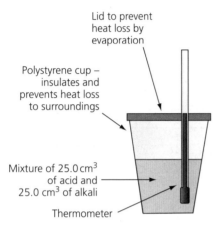

Figure 71 Apparatus for determining enthalpy of neutralisation

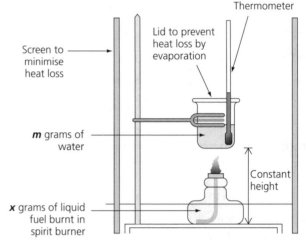

Figure 72 Apparatus for determining enthalpy of combustion

Determining enthalpy of combustion

Figure 72 shows the apparatus used to determine the enthalpy of combustion.

1 Measure the mass of liquid fuel used (this could be the change in mass of the spirit burner or a mass could be calculated from a volume of fuel burnt together with its density).

2 Measure the initial temperature of a known volume of water in a small beaker.

3 Allow the fuel in the spirit burner to burn and to heat the water.

4 Measure the highest temperature reached and calculate the temperature change (ΔT).

5 Calculate the energy change using $q = mc\Delta T$.

6 Calculate the energy change for 1 mol of fuel by dividing by the number of moles of fuel burnt and then divide by 1000 to obtain a value in $kJ\,mol^{-1}$.

Typical calculations

Worked example 1

$200\,cm^3$ of water were heated by burning ethanol in a spirit burner. The following mass measurements were recorded:

Mass of spirit burner and ethanol (before burning) = 58.25 g

Mass of spirit burner and ethanol (after burning) = 57.62 g

The initial temperature of the water was 20.7°C and the highest temperature recorded was 41.0°C. The specific heat capacity of water is $4.2\,J\,g^{-1}\,°C^{-1}$. Calculate the standard enthalpy of combustion of ethanol. →

Exam tip

When enthalpy values are calculated from experimental data, the values may be less than expected, as some heat is lost to the surroundings and some is also lost by evaporation.

Answer

$\Delta T = 41.0 - 20.7 = 20.3°C$

$m = 200\,g$ (volume of water = $200\,cm^3$, so mass of water = $200\,g$)

$c = 4.2\,J\,g^{-1}\,°C^{-1}$

$q = mc\Delta T = 200 \times 4.2 \times 20.3 = 17\,052\,J$

Mass of ethanol burnt = $58.25 - 57.62 = 0.63\,g$

Moles of ethanol burnt = $\dfrac{0.63}{46} = 0.0137\,mol$

q per mol of ethanol = $\dfrac{17\,052}{0.0137} = 1\,244\,672\,J\,mol^{-1}$

Standard enthalpy of combustion = $\dfrac{1\,244\,672}{1000} = -1244.672\,kJ\,mol^{-1}$

Worked example 2

$25.0\,cm^3$ of $2\,mol\,dm^{-3}$ hydrochloric acid and $25.0\,cm^3$ of $2\,mol\,dm^{-3}$ sodium hydroxide solution were mixed in a polystyrene cup. The temperature change recorded was $13.1°C$. Assuming the solution has a specific heat capacity of $4.2\,J\,g^{-1}\,°C^{-1}$ and that the density of the solution is $1\,g\,cm^{-3}$, calculate the standard enthalpy of neutralisation.

Answer

$\Delta T = 13.1°C \qquad m = 50\,g\ (25 + 25) \qquad c = 4.2\,J\,g^{-1}\,°C^{-1}$

$q = mc\Delta T = 50 \times 4.2 \times 13.1 = 2751\,J$

Moles of NaOH = $\dfrac{25 \times 2}{1000} = 0.05\,mol$

Moles of HCl = $\dfrac{25 \times 2}{1000} = 0.05\,mol$

$NaOH + HCl \rightarrow NaCl + H_2O$

$0.05\,mol$ of NaOH react completely with $0.05\,mol$ of HCl to form $0.05\,mol$ of water

q per mol of water formed = $\dfrac{2751}{0.05} = 55\,020\,J$

Standard enthalpy change of neutralisation = $\dfrac{55\,020}{1000} = -55.02\,kJ\,mol^{-1}$.

Exam tip

The density of water is $1\,g\,cm^{-3}$ so $100\,cm^3$ of water = $100\,g$.

Exam tip

The units of standard enthalpy changes are $kJ\,mol^{-1}$. At the end of the calculation, the negative sign is included before the value because the reaction is exothermic. This is a common omission and can cost you a mark. If you calculate an enthalpy change from experimental data, always think about whether the reaction is exothermic or endothermic and place the appropriate sign in front of the value.

Knowledge check 18

In the expression, In $q = mc\Delta T$, what do m, c and ΔT represent? State their units.

Hess's law

Hess's law is a statement of the **principle of conservation of energy**. Hess's law can be used to calculate enthalpy changes for chemical reactions from the enthalpy changes of other reactions. This is useful because it enables theoretical enthalpy changes to be determined for some reactions that cannot be carried out practically.

Worked example 1

Given the standard enthalpy changes of combustion in Table 17, calculate the standard enthalpy of formation of propanone, $C_3H_6O(l)$.

Write equations that represent the enthalpy changes of combustion given in Table 17:

Table 17

	$\Delta_c H^\ominus$ /kJ mol^{-1}
C(s)	−394
H_2(g)	−286
C_3H_6O(l)	−1821

Equation 1: $C(s) + O_2(g) \rightarrow CO_2(g)$ $\qquad \Delta_c H^\ominus = -394 \, kJ \, mol^{-1}$

Equation 2: $H_2(g) + \frac{1}{2}O_2(g) \rightarrow H_2O(l)$ $\qquad \Delta_c H^\ominus = -286 \, kJ \, mol^{-1}$

Equation 3: $C_3H_6O(l) + 4O_2(g) \rightarrow 3CO_2(g) + 3H_2O(l)$ $\quad \Delta_c H^\ominus = -1821 \, kJ \, mol^{-1}$

There are several ways to approach this question. All are acceptable.

Solving by equations

Write the equation for the reaction — in this case the enthalpy of formation of propanone.

Main equation: $3C(s) + 3H_2(g) + \frac{1}{2}O_2(g) \rightarrow C_3H_6O(l)$

You have to try to get from reactants to products in this main equation using Equations 1–3 above. The reactants in this equation — C(s) and H_2(g) — can be burnt to CO_2(g) and H_2O(l) and then the combustion of propanone reversed theoretically to re-form propanone.

3 mol of C(s) react in the main equation, so Equation 1 is multiplied by 3 and becomes:

$3C(s) + 3O_2(g) \rightarrow 3CO_2(g)$ $\qquad \Delta H = 3(-394) \, kJ \, mol^{-1}$

3 mol of H_2(g) react in the main equation, so Equation 2 is multiplied by 3 and becomes:

$3H_2(g) + 1\frac{1}{2}O_2(g) \rightarrow 3H_2O(l)$ $\qquad \Delta H = 3(-286) \, kJ \, mol^{-1}$

1 mol of propanone is formed in the main equation, so Equation 3 should be reversed:

$3CO_2(g) + 3H_2O(l) \rightarrow C_3H_6O(l) + 4O_2(g)$ $\quad \Delta H = +1821 \, kJ \, mol^{-1}$

The sign is changed because the equation is reversed.

The **principle of conservation of energy** states that energy cannot be created or destroyed, but it can change from one form to another.

Hess's law states that the enthalpy change for a reaction is independent of the route taken, provided the initial and final conditions are the same.

Knowledge check 19

State Hess's law.

Exam tip

Reversing an enthalpy change simply changes the sign of the value. All standard enthalpy changes of combustion are per mole of the substance being burnt, so if 3 mol are being burnt the enthalpy change must be multiplied by 3.

If we add these equations together and cancel down the moles of any substance which appears on both sides of the equation, it should result in the main equation and the total of the enthalpy changes should give the enthalpy change for the main equation (Figure 73).

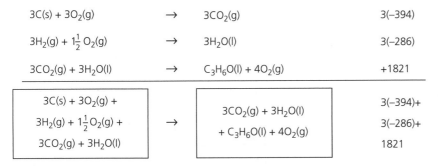

Cancel the moles of substances that appear on both sides of the equation.

$$3C(s) + 3H_2(g) + \tfrac{1}{2}O_2(g) \rightarrow C_3H_6O(l) \qquad -219\,kJ\,mol^{-1}$$

Figure 73 Solving by equations

Solving by Hess's law diagram

As before, start by writing the main equation for the enthalpy change you are to calculate and then write equations for the standard enthalpy changes you have been given to answer the question.

Main equation:

$$3C(s) + 3H_2(g) + \tfrac{1}{2}O_2(g) \rightarrow C_3H_6O(l)$$

Equations for given enthalpy changes:

$$C(s) + O_2(g) \rightarrow CO_2(g) \qquad \Delta_c H^\ominus = -394\,kJ\,mol^{-1}$$

$$H_2(g) + \tfrac{1}{2}O_2(g) \rightarrow H_2O(l) \qquad \Delta_c H^\ominus = -286\,kJ\,mol^{-1}$$

$$C_3H_6O(l) + 4O_2(g) \rightarrow 3CO_2(g) + 3H_2O(l) \qquad \Delta_c H^\ominus = -1821\,kJ\,mol^{-1}$$

The substances which do not appear in the main equation but do appear in the equations for the given enthalpy changes are $CO_2(g)$ and $H_2O(l)$. These are the link substances which will allow you to draw a Hess's law diagram (Figure 74).

- Below the main equation put $3CO_2(g)$ and $3H_2O(l)$.
- Draw arrows for the enthalpy changes given in the question in the direction of the changes.
- Write ΔH values on the arrows and remember to multiply by the number of moles of substance burnt.

The final calculation is from the reactants $3C(s)$, $3H_2(g)$ and $\tfrac{1}{2}O_2(g)$ to the products in the box and then from the box to propanone $C_3H_6O(l)$. The oxygen will balance as the arrows to the box all represent combustion reactions.

→

> **Exam tip**
>
> Use '+ number of moles × (standard enthalpy change)' when following the direction of an arrow and '– number of moles × (standard enthalpy change)' when going against the direction of the arrow. For +3(–394), + is used following the direction of the arrow, 3 is used as 3 mol of carbon are burnt, and –394 is the standard enthalpy of combustion of carbon. Similarly with +3(–286) for H_2. For –(–1821), – is used (against the direction of the arrow), there is no number as 1 mol of propanone is burnt, and –1821 is the standard enthalpy of combustion of propanone.

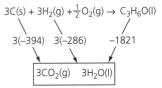

Figure 74 Solution by Hess's law diagram

The standard enthalpy of formation of propanone (main equation) can be calculated by going from the reactants to $CO_2(g)$ and $H_2O(l)$ and then reversing the combustion of propanone.

$$\Delta_f H^{\ominus}(C_3H_6O(l)) = + 3(-394) + 3(-286) - (-1821)$$

$$= -1182 - 858 + 1821 = -219 \, kJ \, mol^{-1}$$

Worked example 2

Calculate the standard enthalpy of combustion of ethane from the standard enthalpy changes of formation given in Table 18.

Answer

Solving by equations

Main equation:

$$C_2H_6(g) + 3\tfrac{1}{2}O_2(g) \rightarrow 2CO_2(g) + 3H_2O(l)$$

Table 18

	$\Delta_f H^{\ominus}/kJ\,mol^{-1}$
$CO_2(g)$	−394
$H_2O(l)$	−286
$C_2H_6(g)$	−84.6

Equations for given enthalpy changes:

    Equation 1: $C(s) + O_2(g) \rightarrow CO_2(g)$ $\Delta_f H^{\ominus} = -394 \, kJ \, mol^{-1}$

2 mol of $CO_2(g)$ are formed in the main equation, so Equation 1 will be multiplied by 2.

    Equation 2: $H_2(g) + \tfrac{1}{2}O_2(g) \rightarrow H_2O(l)$ $\Delta_f H^{\ominus} = -286 \, kJ \, mol^{-1}$

3 mol of $H_2O(l)$ are formed in the main equation, so Equation 2 will be multiplied by 3.

    Equation 3: $2C(s) + 3H_2(g) \rightarrow C_2H_6(g)$ $\Delta_f H^{\ominus} = -84.6 \, kJ \, mol^{-1}$

1 mol of $C_2H_6(g)$ reacts in the main equation, so Equation 3 will be reversed.

The method is summarised in Figure 75.

$2C(s) + 2O_2(g)$	$\rightarrow$	$2CO_2(g)$	$2(-394)$
$3H_2(g) + 1\tfrac{1}{2}O_2(g)$	$\rightarrow$	$3H_2O(l)$	$3(-286)$
$C_2H_6(g)$	$\rightarrow$	$2C(s) + 3H_2(g)$	$+84.6$
$2C(s) + 2O_2(g) +$ $3H_2(g) + 1\tfrac{1}{2}O_2(g) +$ $C_2H_6(g)$	$\rightarrow$	$2CO_2(g) + 3H_2O(l)$ $+ 2C(s) + 3H_2(g)$	$2(-394) +$ $3(-286) +$ 84.6
$C_2H_6(g) + 3\tfrac{1}{2}O_2(g)$	$\rightarrow$	$2CO_2(g) + 3H_2O(l)$	$-1561.4 \, kJ \, mol^{-1}$

Figure 75 Solving by equations

➡

Exam tip

When given standard enthalpy changes of combustion, the enthalpy change of the reaction may be calculated as $\Delta H = \Sigma \Delta_c H(\text{reactants}) - \Sigma \Delta_c H(\text{products})$. $\Sigma \Delta_c H$ (reactants) $= 3(-394) + 3(-286) = -2040$ and $\Sigma \Delta_c H(\text{products}) = -1821$. This gives $-2040 - (-1821) = -219 \, kJ \, mol^{-1}$

Solving by Hess's law diagram

From the equation for the given enthalpy changes the substances which do not appear in the main equation are C(s) and $H_2(g)$. This cycle can be thought of as ethane being converted to its elements in their standard states and the elements reacting with oxygen to form $CO_2(g)$ and $H_2O(l)$. See Figure 76.

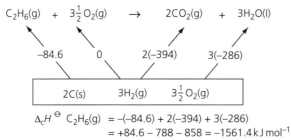

$$\Delta_c H^{\ominus}\ C_2H_6(g) = -(-84.6) + 2(-394) + 3(-286)$$
$$= +84.6 - 788 - 858 = -1561.4\,kJ\,mol^{-1}$$

Figure 76 Solving by Hess's law diagram

Worked example 3

Vanadium metal can be obtained from ore using calcium metal. The reaction is represented by the equation:

$$V_2O_5(s) + 5Ca(s) \rightarrow 2V(l) + 5CaO(s)$$

a Use the standard enthalpies of formation in Table 19 to calculate a value for the standard enthalpy change of this reaction.

b Explain why the standard enthalpy of formation of Ca(s) is zero but the standard enthalpy of formation of V(l) is not zero.

Answer

Solving by equations

Main equation:

$$V_2O_5(s) + 5Ca(s) \rightarrow 2V(l) + 5CaO(s)$$

Equations for given enthalpy changes of formation:

Equation 1: $2V(s) + 2\frac{1}{2}O_2(g) \rightarrow V_2O_5(s)$ $\Delta_f H^{\ominus} = -1551\,kJ\,mol^{-1}$

1 mol of $V_2O_5(s)$ reacts in the main equation, so Equation 1 will be reversed.

Equation 2: $V(s) \rightarrow V(l)$ $\Delta_f H^{\ominus} = +23\,kJ\,mol^{-1}$

2 mol of V(l) are formed in the main equation, so Equation 2 will be multiplied by 2.

Equation 3: $Ca(s) + \frac{1}{2}O_2(g) \rightarrow CaO(s)$ $\Delta_f H^{\ominus} = -635\,kJ\,mol^{-1}$

5 mol of CaO(s) are formed in the main equation, so Equation 3 will be multiplied by 5.

The method is summarised in Figure 77.

Table 19

	$\Delta_f H^{\ominus}/kJ\,mol^{-1}$
$V_2O_5(s)$	−1551
Ca(s)	0
V(l)	+23
CaO(s)	−635

$V_2O_5(s)$	$\rightarrow$	$2V(s) + 2\frac{1}{2}O_2(g)$	$+1551$
$2V(s)$	$\rightarrow$	$2V(l)$	$2(+23)$
$5Ca(s) + 2\frac{1}{2}O_2(g)$	$\rightarrow$	$5CaO(s)$	$5(-635)$
$V_2O_5(s) +$ $2V(s) +$ $5Ca(s) + 2\frac{1}{2}O_2(g)$	$\rightarrow$	$2V(s) + 2\frac{1}{2}O_2(g) +$ $2V(l) + 5CaO(s)$	$1551 +$ $2(23) +$ $5(-635)$
$V_2O_5(s) + 5Ca(s)$	$\rightarrow$	$2V(l) + 5CaO(s)$	$-1578\,kJ\,mol^{-1}$

Figure 77 Solving by equations

Solving by Hess's law diagram

From the equation for the given enthalpy changes, the substances which do not appear in the main equation are $V(s)$ and $O_2(g)$; $Ca(s)$ will also appear in the box to link $Ca(s)$ with $CaO(s)$. This cycle can be thought of as V_2O_5 being converted to its elements and the elements reacting to form the products. See Figure 78.

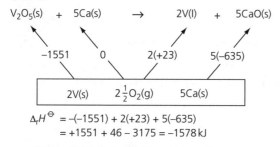

$$\Delta_rH^{\ominus} = -(-1551) + 2(+23) + 5(-635)$$
$$= +1551 + 46 - 3175 = -1578\,kJ$$

Figure 78 Solving by Hess's law diagram

When given standard enthalpy changes of formation, the enthalpy change of the reaction you are looking for is simply calculated from:

$$\Delta H = \Sigma\Delta_fH(\text{products}) - \Sigma\Delta_fH(\text{reactants})$$

In this reaction, $\Sigma\Delta_fH(\text{products}) = -1551$ and $\Sigma\Delta_fH(\text{reactants}) = 2(+23) + 5(-635)$ $= -3129$. This gives:

$$-3129 - (-1551) = -1578\,kJ$$

b $Ca(s)$ is an element in its standard state whereas $V(l)$ is an element but it is not in its standard state under standard conditions. $V(s)$ is an element in its standard state.

Exam tip

The $Ca(s)$ is an element in its standard state and so its enthalpy of formation value is zero.

Knowledge check 20

The enthalpies of formation of calcium oxide, water and calcium hydroxide are -636, -286 and $-987\,kJ\,mol^{-1}$ respectively. Calculate the enthalpy change for the reaction $CaO(s) + H_2O(l) \rightarrow Ca(OH)_2(s)$.

Average bond enthalpy

The **average bond enthalpy** is a measure of the energy required to break 1 mol of a covalent bond averaged across many compounds containing the bond. It is measured in kJ mol⁻¹ (the 'per mole' is per mole of the covalent bond). For example, the average bond enthalpy of C–H is 413 kJ mol⁻¹.

Remember that bond breaking is endothermic and bond making is exothermic.

- Breaking 1 mol of C–H $\Delta H = +413$ kJ mol⁻¹   (positive as endothermic)
- Making 1 mol of C–H $\Delta H = -413$ kJ mol⁻¹   (negative as exothermic)

Worked example 1

Determine the standard enthalpy of combustion of ethene (C_2H_4) using the average bond enthalpy values given in Table 20.

Answer

The equation for the reaction should be written as a balanced symbol equation. (Remember that to represent a standard enthalpy change of combustion the equation should be per mole of the fuel (ethene) burnt.)

Table 20

Bond	Average bond enthalpy/kJ mol⁻¹
C—H	413
C=C	611
O—H	464
O=O	497
C=O	803

$$C_2H_4(g) + 3O_2(g) \rightarrow 2CO_2(g) + 2H_2O(g) \quad \text{(per mole of } C_2H_4\text{)}$$

Then draw a structural equation showing all the covalent bonds:

Calculate the energy required to break the bonds in the reactants:

1 C=C bond = 611 kJ mol⁻¹

4 C—H bond = 4(413) = 1652 kJ mol⁻¹

3 O=O bond = 3(497) = 1491 kJ mol⁻¹

Total energy required for bonds broken = 3754 kJ mol⁻¹

Calculate the energy released when bonds form in the products:

4 C=O bonds = 4(803) = 3212 kJ mol⁻¹

4 O—H bonds = 4(464) = 1856 kJ mol⁻¹

Total energy released for bonds made = 5068 kJ mol⁻¹

ΔH = sum of mean bond enthalpies of bonds broken − sum of mean bond enthalpies of bonds made

Overall enthalpy change for the reaction = +3754 − 5068 = −1314 kJ mol⁻¹

The average bond enthalpy is the energy required to break 1 mol of a given bond averaged over many compounds.

Exam tip

The answers to calculations involving average bond enthalpy values often differ from the actual values because average bond enthalpy values are not specific to the molecules in the reaction. A common question is why the enthalpy change calculated from average bond enthalpy values is different to the known value.

Exam tip

The overall enthalpy change is the energy required to break bonds in reactants minus the energy released when bonds form in products. The energy required to break the bonds is positive (+) as it is endothermic and the energy released when bonds form is negative (–) as it is exothermic.

Content Guidance

The standard enthalpy of combustion of ethanol, CH_3CH_2OH, is $-1367\,kJ\,mol^{-1}$.

Calculate a value for the bond enthalpy of the C–O bond in ethanol using the average bond enthalpies given in Table 21.

Answer

Equation for combustion of ethanol:

$$CH_3CH_2OH(g) + 3O_2(g) \rightarrow 2CO_2(g) + 3H_2O(g) \quad \text{(per mole of } CH_3CH_2OH)$$

Calculate the energy required to break the bonds in the reactants:

1 mol of C—C bonds $= 348\,kJ\,mol^{-1}$

5 mol of C—H bonds $= 5(413) = 2065\,kJ\,mol^{-1}$

1 mol of C—O bonds $= x\,kJ\,mol^{-1}$, where x is unknown

1 mol of O—H bonds $= 464\,kJ\,mol^{-1}$

3 mol of O=O bonds $= 3(497) = 1491\,kJ\,mol^{-1}$

Total energy required for bonds broken $= 4368 + x\,kJ\,mol^{-1}$

Calculate the energy released when bonds form in the products:

4 C=O bonds $= 4(803) = 3212\,kJ\,mol^{-1}$

6 O—H bonds $= 6(464) = 2784\,kJ\,mol^{-1}$

Total energy released for bonds made $= 5996\,kJ\,mol^{-1}$

$\Delta H =$ sum of mean bond enthalpies of bonds broken − sum of mean bond enthalpies of bonds made

$\Delta H = -1367 = +4368 + x - 5996$

$x = -1367 - 4368 + 5996 = 261\,kJ\,mol^{-1}$

The quoted average bond enthalpy value is $360\,kJ\,mol^{-1}$ for the C–O bond, but this bond occurs in many different molecules and this calculation used the standard enthalpy of combustion of ethanol, making the calculation specific to this molecule. Inconsistencies in values obtained in calculations of this sort are common with the use of average bond enthalpies.

Table 21

Bond	Average bond enthalpy/kJ mol⁻¹
C—H	413
C—C	348
O—H	464
O=O	497
C=O	803

Exam tip

When using average bond enthalpies, all substances are assumed to be in the gaseous state so $CH_3CH_2OH(g)$ and $H_2O(g)$ appear in the equation. This assumes no intermolecular forces. This can be another reason why the standard enthalpy change (where $H_2O(l)$ is used) differs from the value obtained using average bond enthalpies.

Summary

- ΔH is the change in enthalpy in a system (reaction). When ΔH is positive the reaction is endothermic and when ΔH is negative the reaction is exothermic.
- An enthalpy-level diagram shows enthalpy against progress of reaction (reaction coordinate).
- Standard enthalpy changes of combustion, neutralisation, formation and reaction are all under standard conditions of 298K (25°C) and 100kPa.
- Temperature changes from experiments can be used to calculate enthalpy changes using $q = mc\Delta T$.
- Hess's law allows calculation of an enthalpy change using other standard enthalpy changes.
- Bond breaking is endothermic and bond making is exothermic.
- Average bond enthalpy values can also be used to calculate an enthalpy change using the energy of the covalent bonds broken and made in the reaction.
- An average bond enthalpy may be calculated from the enthalpy change for the reaction.

■ Kinetics

Chemical kinetics is the study of rates of reaction.

- A catalyst provides an alternative reaction pathway of lower activation energy.
  - A **homogeneous catalyst** is a catalyst that is in the same state as the reactants and products and provides a pathway of lower activation energy by forming an intermediate.
  - A **heterogeneous catalyst** is a catalyst that is in a different state to the reactants and products and works via chemisorption and provides a pathway of lower activation energy.
- **Chemisorption** is the process whereby reactant molecules are adsorbed onto the surface of the catalyst, bonds are weakened, reactant molecules are held in a more favourable orientation for reaction and product molecules are desorbed from the catalyst.

Measuring rate of reaction

Many chemical reactions produce a gas. When a gas is produced, the rate of reaction may be measured using several methods. The main methods used are:

- gas volume against time
- mass against time

There are other methods used to measure the rate of a chemical reaction but these two methods will provide quantitative data.

A gas syringe or a burette/measuring cylinder full of water inverted over water may be used to measure gas volume. A container on a balance may be used to measure the mass of the reaction mixture. Graphs of gas volume against time and mass against time may be plotted. Changes in conditions may alter the shape of the graph. A faster reaction will cause a more rapid change in gas volume and mass. This was studied more extensively at GCSE and will be extended at A2.

Factors affecting rate of reaction

Reaction rate depends on the number of successful collisions between reacting particles in a given period of time. The number of collisions can be affected by

The **reaction rate** is the change in the concentration (amount) of a reactant or product with respect to time.

A **catalyst** is a substance that increases the speed of a chemical reaction but does not get used up.

Activation energy is the minimum amount of energy required for a reaction to occur.

Knowledge check 21

Explain how a catalyst increases the rate of reaction.

changes in **temperature**, **pressure**, **concentration** and the **presence of a catalyst**. A successful collision is a collision in which the reacting particles have enough energy to react (i.e. they possess at least the activation energy).

Temperature

Increasing temperature increases the energy of reacting particles, which leads to an increase in the number of collisions. This causes an increase in the number of successful collisions in a given period of time, which increases the rate of reaction.

Pressure

Increasing the pressure pushes the reacting particles closer together, which increases the number of collisions. This causes an increase in the number of successful collisions in a given period of time, which increases the rate of reaction.

Concentration

Increasing the concentration of the reactant(s) increases the number of reacting particles, which leads to an increased number of collisions. This causes an increase in the number of successful collisions in a given period of time, which increases the rate of reaction.

Presence of a catalyst

A catalyst provides an alternative reaction pathway of lower activation energy, which increases the number of successful collisions and so increases the rate of reaction.

Enthalpy-level diagrams

In the section on energetics (see page 65), we examined enthalpy-profile diagrams for exothermic and endothermic reactions. We now have to include the reaction pathway in the diagram and be able to recognise and/or label the activation energy. Figure 79 is an enthalpy-level diagram for an exothermic reaction.

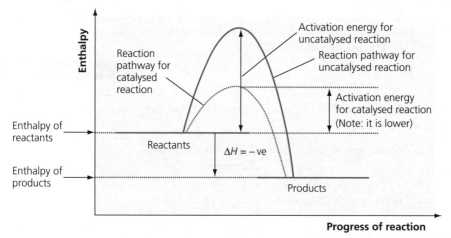

Figure 79 Enthalpy-level diagram for an exothermic reaction

The same is true for an endothermic reaction. The catalyst simply provides an alternative reaction pathway of lower activation energy.

Exam tip

You may be asked to sketch an enthalpy-profile diagram. Do not confuse this with a Maxwell–Boltzmann distribution.

Maxwell–Boltzmann distribution

A Maxwell–Boltzmann distribution is also called a distribution of molecular energies. It is a plot of the number of gaseous molecules against the energy they possess. It is a roughly normal distribution that is asymptotic to the horizontal axis (gets closer and closer but never touches the axis) at higher energy.

Features of a Maxwell–Boltzmann distribution

The features of a Maxwell–Boltzmann distribution are shown in Figure 80.

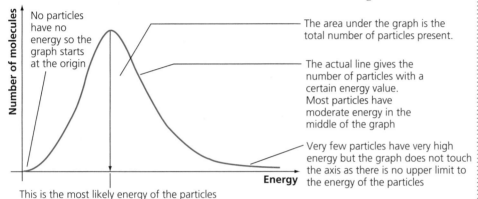

No particles have no energy so the graph starts at the origin

The area under the graph is the total number of particles present.

The actual line gives the number of particles with a certain energy value. Most particles have moderate energy in the middle of the graph

Very few particles have very high energy but the graph does not touch the axis as there is no upper limit to the energy of the particles

This is the most likely energy of the particles

Figure 80 A Maxwell–Boltzmann distribution

Knowledge check 22

What labels are placed on the vertical and horizontal axes of a Maxwell Boltzmann distribution?

Activation energy on a Maxwell–Boltzmann distribution

The activation energy (E_a) is the minimum energy that gaseous reactant molecules must possess to undergo a reaction. It is the energy value of the top of the pathway hump in an enthalpy-profile diagram. If the Maxwell–Boltzmann distribution represents the energy of the reactant molecules there will be an energy value on the x-axis that is the activation energy (Figure 81). The shaded area in Figure 81 represents all the reactant molecules that have energy above the activation energy so they have enough energy to react.

Remember that a catalyst increases the rate of the reaction by providing an alternative reaction pathway of **lower activation energy**. The effect of adding a catalyst is shown in Figure 82. The shaded area in Figure 82 represents all the reactant molecules that have energy above the activation energy. This area is larger than that in Figure 81 (with no catalyst) so more molecules have enough energy to react. This leads to a faster reaction rate.

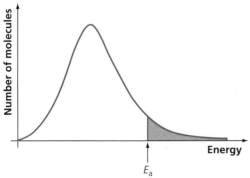

Figure 81 Activation energy

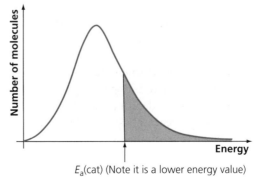

E_a(cat) (Note it is a lower energy value)

Figure 82 Effect of a catalyst on activation energy

Maxwell–Boltzmann distribution at different temperatures

When the temperature is increased, the energy of the gaseous reactant molecules is increased. This changes the shape of the Maxwell–Boltzmann distribution. Figure 83 shows Maxwell–Boltzmann distributions at 300 K and 350 K (the Kelvin temperature scale is often used).

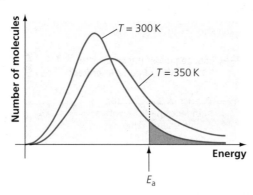

Figure 83 Maxwell–Boltzmann distributions
at two temperatures

Comparing the area under the curve above the activation energy, it is clear that at the higher temperature there are more reactant molecules with enough energy to react. This explains why there is a higher rate of reaction at a higher temperature.

Exam tip

The total area under the graph must be the same because the concentrations of the reactant molecules are all the same. This means that as the energy of the particles increases the curve becomes lower but more spread out. You will often be asked to sketch a curve like this at a lower or higher temperature. Lower temperature curves are narrower and have a higher peak; higher temperature curves are wider with a lower peak.

Summary

- The activation energy is the minimum energy that the reactant molecules need for a reaction to occur. It can be represented on an enthalpy-profile diagram or on a Maxwell–Boltzmann distribution.
- A Maxwell–Boltzmann distribution shows the distribution of molecular energies of the reactant particles. The graph changes shape with changes in temperature.
- A catalyst increases the rate of reaction by providing an alternative reaction pathway of lower activation energy.

■ Equilibrium

Many reactions are **reversible**. The reactions start with the reactants and as the products form, some products break down to form reactants again. **Equilibrium** is achieved when the amount of reactants and products remains constant. An equilibrium is described as **dynamic** when the rate of the forward reaction is equal to the rate of the reverse reaction, resulting in the amount of reactants/products remaining constant. Reactions are described as **homogeneous** or **heterogeneous**, depending on the states of the reactants and products. In a homogeneous reaction, all the reactants and products are in the same physical state but in a heterogeneous reaction all the reactants and products are *not* in the same physical state.

A **reversible reaction** is a reaction that goes in both forward and reverse directions.

Equilibrium is a reversible reaction in which the amount of each reactant/product remains constant.

A **dynamic equilibrium** is a reaction in which the amounts of each reactant/product remain constant because the rate of the forward reaction is equal to the rate of the reverse reaction.

A **homogeneous reaction** is a reaction in which all reactants and products are in the same physical state.

A **heterogeneous reaction** is a reaction in which all the reactants and products are not in the same physical state.

Position of equilibrium

The position of equilibrium in a reversible reaction is a measure of how far the reaction has proceeded to the right (towards the products) or has remained to the left (towards the reactants).

- When the equilibrium lies to the left, it means that little product is formed (reactants predominate).
- When the equilibrium lies to the right, it means that little reactant remains (products predominate).

Factors affecting the position of equilibrium

Several factors affect the position of equilibrium. These include concentration, temperature and pressure in homogeneous gaseous reactions.

Concentration

- An increase in the concentration of a reactant moves the position of equilibrium to the right.
- An increase in the concentration of a product moves the position of equilibrium to the left.
- A decrease in the concentration or removal of a reactant moves the position of equilibrium to the left.
- A decrease in the concentration or removal of a product moves the position of equilibrium to the right.

Temperature

Changes in temperature affect the position of equilibrium based on whether the reaction is exothermic or endothermic.

- If the forward reaction is endothermic (ΔH is positive):
  - An increase in temperature moves the position of equilibrium in the direction of the forward (endothermic) reaction. Position of equilibrium moves to the right to oppose the change (higher yield of products).
  - A decrease in temperature moves the position of equilibrium in the direction of the reverse (exothermic) reaction. Position of equilibrium moves to the left to oppose the change (lower yield of products).

Exam tip

A reversible reaction is shown with a ⇌ as opposed to a traditional →. The double arrowhead indicates that the reaction is reversible.

Knowledge check 23

What is meant by a dynamic equilibrium?

- If the forward reaction is exothermic (ΔH is negative):
  - An increase in temperature moves the position of equilibrium in the direction of the reverse (endothermic) reaction. Position of equilibrium moves to the left to oppose the change (lower yield of products).
  - A decrease in temperature moves the position of equilibrium in the direction of the forward (exothermic) reaction. Position of equilibrium moves to the right to oppose the change (higher yield of products).

Pressure

A change in pressure is only applied at A-level to a homogeneous gas system in which all reactants and products are gases. Count the total number of moles of gas on the reactant side and on the product side, based on the balancing numbers in the equation. If there are the same number of moles of gas on each side, a change in pressure will have no effect on the position of equilibrium.

- If left-hand side (reactants) has fewer gas moles:
  - An increase in pressure moves the position of equilibrium in the direction of the equation side with fewer gas moles. Position of equilibrium moves to the left to oppose the change (lower yield of products).
  - A decrease in pressure moves the position of equilibrium in the direction of the equation side with more gas moles. Position of equilibrium moves to the right to oppose the change (higher yield of products).
- If the right-hand side (products) has fewer gas moles:
  - An increase in pressure moves the position of equilibrium in the direction of the equation side with fewer gas moles. Position of equilibrium moves to the right to oppose the change (higher yield of products).
  - A decrease in pressure moves the position of equilibrium in the direction of the equation side with more gas moles. Position of equilibrium moves to the left to oppose the change (lower yield of products).

High pressure is expensive to apply and requires thick-walled vessels to contain it. There are also safety issues with high pressures because there is an increased risk of explosion.

Presence of a catalyst

The presence of a catalyst has no effect on the position of equilibrium, but allows equilibrium to be attained more quickly.

Equilibrium constant, K_c

- For the general reaction:

$$aA + bB \rightleftharpoons cC + dD$$

$$K_c = \frac{[C]^c[D]^d}{[A]^a[B]^b}$$

where [C] represents the concentration of C in $mol\,dm^{-3}$ in the equilibrium mixture and c is the balancing number for C in the equation for the reaction. The same applies to A, B and D.

Exam tip

A decrease in volume has the same effect as an increase in pressure.

Exam tip

If a question asks you to state and explain how changing an external factor such as temperature, concentration or pressure affects the yield of a particular substance in the reaction, make sure you explain this in terms of how the position of equilibrium moves and why it moves but also whether the yield increases or decreases.

Exam tip

Square brackets around a substance indicate its concentration measured in $mol\,dm^{-3}$. Make sure you only use square brackets in a K_c expression.

- The concentrations of all products at equilibrium are on the top line of the expression raised to the power of their balancing numbers and the concentrations of all reactants at equilibrium are on the bottom line of the expression, again raised to the power of their balancing numbers.
- Concentration is often calculated as the number of moles of a reactant or product divided by the volume (most often in dm^3).
- Units of $K_c = \dfrac{(mol\,dm^{-3})^{(c+d)}}{(mol\,dm^{-3})^{(a+b)}}$.
- The units are given in terms of concentration in $mol\,dm^{-3}$, but the overall power depends on the balancing numbers in the equation for the reaction.
- A high K_c value would indicate that the position of equilibrium is to the right-hand side.
- A low K_c value would indicate that the position of equilibrium is to the left-hand side.
- The higher the value of K_c the further the position of equilibrium is to the right.

Writing K_c expressions and calculation of units of K_c

A common question is to write an expression for the equilibrium constant, K_c and to deduce the units of K_c.

For the general reaction:

$$2A + B \rightleftharpoons 2C$$

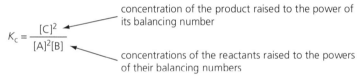

concentration of the product raised to the power of its balancing number

$$K_c = \frac{[C]^2}{[A]^2[B]}$$

concentrations of the reactants raised to the powers of their balancing numbers

Figure 84 Expression for K_c

The [B] does not need a power as its balancing number is 1 and [B] is the same as $[B]^1$.

The units are worked out using $mol\,dm^{-3}$ as the unit of all the concentrations.

$$\text{units of } K_c = \frac{(mol\,dm^{-3})^2}{(mol\,dm^{-3})^2(mol\,dm^{-3})} = \frac{(mol\,dm^{-3})^2}{(mol\,dm^{-3})^3} = (mol\,dm^{-3})^{-1} = mol^{-1}\,dm^3$$

Worked example

Write an expression for K_c for the reaction:

$$N_2(g) + 3H_2(g) \rightleftharpoons 2NH_3(g)$$

and calculate its units.

Answer

$$K_c = \frac{[NH_3]^2}{[N_2][H_2]^3}$$

$$\text{Units of } K_c = \frac{(mol\,dm^{-3})^2}{(mol\,dm^{-3})^4} = mol^{-2}\,dm^6$$

Exam tip

Do not put + in the expression. This is a common mistake. The bottom line of the expression is the concentration of A squared multiplied by the concentration of B.

Exam tip

The concentration of a solution or a gas is expressed in $mol\,dm^{-3}$ as it is a measure of the amount of a substance, in moles, present in a volume of $1\,dm^3$.

Remember that when the same term is multiplied their powers are added and if the same term is divided then the powers are subtracted. In the example, concentration^2 divided by concentration^4 = concentration$^{(2-4)}$ = concentration^{-2} = $mol^{-2} dm^6$.

No units of K_c

For the following reaction:

$$2HI \rightleftharpoons H_2 + I_2$$

Equilibrium constant, K_c, has no units.

$$K_c = \frac{[HI]^2}{[H_2][I_2]}$$

Units of $K_c = \frac{(mol dm^{-3})^2}{(mol dm^{-3})^2}$ = no units

If K_c has no units, it is because there are equal numbers of moles on both sides of the equation and they cancel each other out in the K_c expression.

Effect of changes of conditions on K_c

K_c values are only affected by changes in temperature. Changes in concentration or pressure or the presence of a catalyst do not have any effect on the value of K_c provided the temperature remains constant.

Change in temperature

An exothermic reaction is one in which the forward reaction releases energy in the form of heat.

■ Increasing the temperature of an exothermic reaction will move the position of equilibrium in the direction of the reverse endothermic reaction (to absorb heat) and so the position of equilibrium will move to the left and the value of the equilibrium constant will be lower.

■ Increasing the temperature of an endothermic reaction will move the position of equilibrium in the direction of the forward endothermic reaction (to absorb heat) and so the position of equilibrium will move to the right and the value of the equilibrium constant will be greater.

This information can be used to predict whether a reaction is exothermic or endothermic based on the values of the equilibrium constant at different temperatures. If K_c increases as temperature increases then the forward reaction is endothermic since equilibrium is moving to the right as temperature increases. If K_c decreases as temperature increases then the forward reaction is exothermic.

Equilibrium in industrial reactions

Ammonia is produced industrially from nitrogen and hydrogen in the Haber process. The reaction is a homogeneous gaseous equilibrium. The equation and enthalpy change of reaction are:

$$N_2 + 3H_2 \rightleftharpoons 2NH_3 \qquad \Delta H = -92.2 \, kJ$$

Knowledge check 24

$$K_c = \frac{[H_2][Br_2]}{[HBr]^2}$$

Write an equation for the equilibrium for the reaction to which this relates.

Exam tip

For an equilibrium reaction if $K_c > 1$, then there are generally more products than reactants. If $K_c < 1$, there would be generally more reactants than products.

- With an increase in temperature, the position of equilibrium moves to the left (moves the position of equilibrium in the direction of the endothermic reaction) and there is a lower yield of product.
- With an increase in pressure, the position of equilibrium moves to the right (fewer gas moles) and there is a higher yield of product.
- In the presence of iron catalyst, equilibrium is attained more quickly.

Conditions: 450°C —compromise temperature

200 atm — high pressure to favour NH_3 production

iron catalyst — equilibrium attained more quickly

A lower temperature would give a higher yield of ammonia but the rate of reaction is also lower. A **compromise temperature** is used because it is a compromise between achieving a reasonable yield with a fast enough rate of reaction. Iron is a heterogeneous catalyst in this reaction.

In the Contact process for the production of concentrated sulfuric acid, sulfur dioxide reacts with oxygen to form sulfur trioxide. The equation and enthalpy change of reaction are:

$$2SO_2 + O_2 \rightleftharpoons 2SO_3 \quad \Delta H = -197 \, kJ$$

- An increase in temperature causes the position of equilibrium to move to the left (moves the position of equilibrium in the direction of the endothermic reaction) and there is a lower yield of product.
- An increase in pressure causes the position of equilibrium to move to the right (fewer gas moles) and there is a higher yield of product.
- The presence of a vanadium(v) oxide catalyst results in equilibrium being attained more quickly.

Conditions: 450°C — compromise temperature

2 atm — slightly increased pressure to favour SO_3 production

V_2O_5 catalyst — equilibrium attained more quickly

In a reaction in which the catalyst is in a different phase (state) from the reactants, the catalyst is described as a **heterogeneous catalyst**.

Knowledge check 25

Name the catalyst used in the Haber process.

Summary

- A dynamic equilibrium is a reversible reaction in which the amount of each reactant/product remains constant and the rate of the forward reaction is equal to the rate of the reverse reaction.
- The position of equilibrium moves to minimise the effect of any external stress placed upon it (such as a change in concentration, temperature or pressure).
- If the concentration of a reactant is increased, the position of equilibrium moves to the right.
- If the temperature is increased and the forward reaction is exothermic, the position of equilibrium moves to the left.

- If the pressure is increased, the position of equilibrium moves to the side with the fewer number of gas moles.
- A catalyst has no effect on the position of equilibrium, but will increase the rate of reaction, so that equilibrium is attained more quickly.
- The magnitude of the equilibrium constant, K_c, relates to the position of equilibrium; a higher value indicates that the position of equilibrium is more to the right-hand side with more product formed.
- K_c is only affected by changes in temperature.
- A compromise temperature is a temperature used in industry to maximise the yield at the fastest rate possible.

Group II elements and their compounds

Group II elements are called s-block elements as their outer shell (highest energy) electrons are in an s subshell. Beryllium is not studied as its chemistry is unusual.

Trends from magnesium to barium

- The atoms of these elements all have an electronic configuration which ends in s^2.

    Mg $1s^2\, 2s^2\, 2p^6\, 3s^2$ or [Ne] $3s^2$

    Ca $1s^2\, 2s^2\, 2p^6\, 3s^2\, 3p^6\, 4s^2$ or [Ar] $4s^2$

    Sr   [Kr] $5s^2$

    Ba   [Xe] $6s^2$

- Atomic radius increases down the group — the outer shell electrons are further from the nucleus.
- First ionisation energy decreases down the group — atomic radius increases and shielding by inner electrons increases.

Reactions of group II elements

In the following examples M represents any of the group II metals in the general equations for the reactions of group II metals with oxygen, water/steam and dilute mineral acids.

Group II elements with oxygen

When heated in air all of the group II elements react with oxygen. When heated in pure oxygen they react more vigorously.

General equation: $2M + O_2 \rightarrow 2MO$

Example: $2Mg + O_2 \rightarrow 2MgO$

Observations:

- Magnesium burns with a white light and produces a white powder.
- Calcium burns with a brick red flame and produces a white powder.
- Strontium burns with a red flame and produces a white powder.
- Barium burns with a green flame and produces a white powder.

Exam tip

The simple compounds of group II elements are white when solid and if they dissolve in water they form colourless solutions. These facts are useful when trying to work out observations for familiar and unfamiliar chemical reactions.

An s-block element is an element that has an atom with highest energy/outer electron in an s subshell (orbital).

Atomic radius is half the distance between the centres of a pair of atoms.

Knowledge check 26

State the trend in first ionisation energy and atomic radius going down group II.

Exam tip

You are expected to recall and apply your knowledge of the reactions of group II elements with oxygen, water/steam and dilute mineral acids. Many of these reactions were studied at GCSE. Remember that reactivity increases down the group and that the equations for the reactions are similar as the metals form simple ions with a 2+ charge. This means that the formulae of the compounds are also similar.

Group II elements with water/steam

Magnesium reacts very slowly with cold water, but does react readily with steam. The other group II elements (Ca, Sr and Ba) react readily with water.

Group II elements with water

General equation: $M + 2H_2O \rightarrow M(OH)_2 + H_2$

Example: $Ca + 2H_2O \rightarrow Ca(OH)_2 + H_2$

Observations:

■ For magnesium: few bubbles produced slowly; metal dulls.

■ For calcium, strontium and barium, heat is released; bubbles are produced; the metal disappears.

■ When calcium reacts with water the solution appears milky (cloudy white) as calcium hydroxide is less soluble in water than either strontium hydroxide or barium hydroxide. Calcium sinks and then rises due to the rapid production of hydrogen which raises the calcium granules.

The apparatus for a group II element reacting with water is shown in Figure 85.

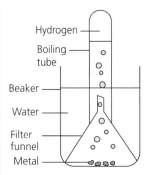

Figure 85 Apparatus for a group II metal reacting with water

Group II elements with steam

General equation: $M + H_2O \rightarrow MO + H_2$

Example: $Mg + H_2O \rightarrow MgO + H_2$

If state symbols are required, the equation would read as follows:

$Mg(s) + H_2O(g) \rightarrow MgO(s) + H_2(g)$

Observations: Magnesium burns with a white light and a white powder is produced.

The apparatus to react a group II element with steam is shown in Figure 86.

The other group II elements react with increasing vigour, burning with a characteristic red flame (for calcium and strontium) and green flame (for barium) and forming a white solid. The reaction with magnesium and steam is the most studied as it is the safest to carry out in the laboratory.

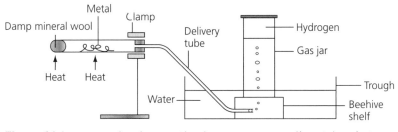

Figure 86 Apparatus for the reaction between a group II metal and steam

Knowledge check 27

Write an equation for the reaction of strontium with water.

Exam tip

Metals react with water to form the metal hydroxide and hydrogen whereas the metals react with steam to form the metal oxide and hydrogen. Make sure you can draw and label the apparatus for the reaction of the group II metals with water and with steam.

Group II elements with dilute mineral acids

All group II metals react with dilute hydrochloric acid and dilute sulfuric acid to produce a salt and hydrogen.

General equations: $M + 2HCl \rightarrow MCl_2 + H_2$

$$M + H_2SO_4 \rightarrow MSO_4 + H_2$$

Example: $Ca + 2HCl \rightarrow CaCl_2 + H_2$

$$Mg + H_2SO_4 \rightarrow MgSO_4 + H_2$$

Observations: heat is released; bubbles are produced; the metal disappears; a colourless solution is formed.

Basic nature of group II oxides

Group II metal oxides are described as bases or basic oxides. **Bases** act as proton (hydrogen ion, H^+) acceptors — they can accept a proton from acids or water.

> **Exam tip**
>
> Oxides of elements are either described as basic (they react with acids), acidic (they react with alkalis/bases), neutral (they react with neither acids nor alkalis/bases) or amphoteric (they react with acids and alkalis/bases). Most metal oxides are basic or amphoteric. If an acidic oxide reacts with water, it forms an acidic solution; and if a basic oxide reacts with water, it forms an alkaline solution. Not all oxides react with water.

Reaction of group II metal oxides with water

General equation: $MO + H_2O \rightarrow M(OH)_2$

Example: $CaO + H_2O \rightarrow Ca(OH)_2$

Observations: heat released; solid expands/crumbles; hisses; colourless solution formed.

In this reaction, the oxide ion O^{2-} accepts a proton (H^+) from water to form the hydroxide ion. The ionic equation for this reaction is:

$$O^{2-} + H^+ \rightarrow OH^-$$

Reaction of group II metal oxides with dilute mineral acids

All group II metal oxides react with dilute mineral acids. The metal oxides are bases. Some of the reactions require warming.

General equations: $MO + 2HCl \rightarrow MCl_2 + H_2O$

$$MO + H_2SO_4 \rightarrow MSO_4 + H_2O$$

$$MO + 2HNO_3 \rightarrow M(NO_3)_2 + H_2O$$

Observations: heat released; solid disappears; colourless solution formed.

> **Exam tip**
>
> Reactions of nitric acid with group II metals are complex as the nitrogen in nitric acid is easily reduced to gaseous products such as nitrogen(II) oxide (NO) also called nitric oxide and nitrogen(IV) oxide (NO_2) also called nitrogen dioxide (brown gas).

> A **base** is a proton acceptor.

> **Exam tip**
>
> Calcium oxide is the metal oxide most often studied in its reaction with water. Calcium oxide, formed from the thermal decomposition of calcium carbonate, is allowed to cool and drops of water are added.

Uses of group II compounds

Magnesium oxide is used in indigestion remedies as it reacts with excess stomach acid. Calcium carbonate is used in toothpaste because it neutralises acid on the teeth which causes dental decay.

Thermal stability of group II carbonates and hydroxides

The thermal stability of a compound is how stable it is to heat and whether it will decompose on heating. The thermal stability of a compound depends on the charge density of the cation and the ability of the cation to polarise and destabilise the anion.

Trend in thermal stability of group II carbonates

Group II carbonates are more thermally stable down the group (Table 22). From $MgCO_3$ to $BaCO_3$ more energy is needed to decompose them.

General equation: $MCO_3 \rightarrow MO + CO_2$

Example: $CaCO_3 \rightarrow CaO + CO_2$

The trend in thermal stability of the group II carbonates can be explained by examining the size of the metal cation (linked to charge density). The carbonate ion can be polarised and destabilised by the larger charge density of a smaller ion such as Mg^{2+}. Ba^{2+} ions have a lower charge density, so the carbonate ion is not as polarised or destabilised by the Ba^{2+} ion. This means that barium carbonate requires more energy to thermally decompose it than the other group II carbonates.

Explaining the trend in thermal stability

- The size of the metal cation *increases* down the group.
- The charge density of the cation *decreases* down the group.
- Going down the group, the cation is less able to polarise and destabilise the carbonate ion.

This means that group II carbonates are more thermally stable as the group is descended.

Trend in thermal stability of group II hydroxides

The thermal stability of the group II hydroxides shows a similar trend (Table 23). Going down the group, the size of the cation increases so the charge density of the cation decreases. Lower charge density cations are less able to polarise and destabilise the hydroxide ion. More energy is required to decompose the group II hydroxides as the group is descended.

General equation: $M(OH)_2 \rightarrow MO + H_2O$

Example: $Ca(OH)_2 \rightarrow CaO + H_2O$

Table 22 Decomposition temperatures of group II carbonates

Compound	Decomposition temperature/°C
$MgCO_3$	540
$CaCO_3$	900
$SrCO_3$	1290
$BaCO_3$	1360

Exam tip

Charge density is the magnitude of charge divided by the volume of the ion. Small ions with a high charge have a higher charge density. For example, Mg^{2+} would have a higher charge density than the larger Ba^{2+}. Both have the same charge.

Knowledge check 28

State the trend in thermal stability of the group II carbonates going down the group.

Table 23

Compound	Decomposition temperature/°C
$Mg(OH)_2$	332
$Ca(OH)_2$	580
$Sr(OH)_2$	710
$Ba(OH)_2$	800

Uses of calcium compounds

Calcium carbonate is the main component of limestone and marble. When limestone is heated in a limekiln it thermally decomposes to form calcium oxide (which is known as quicklime or lime). Carbon in the form of coke is added to the limekiln to burn in the hot air blasted in to provide the heat for the thermal decomposition.

Calcium oxide reacts with water to form calcium hydroxide (which is known as slaked lime). The reaction is vigorous. The calcium oxide hisses and expands and crumbles. If water is added until it is in excess, the solution formed is cloudy. It is called limewater (calcium hydroxide solution). The reaction is:

$$CaO + H_2O \rightarrow Ca(OH)_2$$

Cement and concrete

Cement is a mixture of calcium oxide (or calcium hydroxide) with silica (silicon dioxide), which is mainly sand and aluminium oxide. When water is added, the reaction between the basic calcium oxide and acidic silicon dioxide creates calcium silicate. Aluminium oxide reacts with calcium oxide to form calcium aluminate. As the water evaporates the calcium silicate and calcium aluminate crystallise and form the solid cement. The reactions are:

$$CaO + SiO_2 \rightarrow CaSiO_3 \qquad CaO + Al_2O_3 \rightarrow Ca(AlO_2)_2$$

Aluminium oxide is amphoteric, which means it can react with acids and with bases. When it reacts with bases it forms salts called aluminates.

Concrete is cement with gravel and more sand added to provide solid mass and it is also mixed to aerate the mixture. Concrete is our main building material, used in the construction of many buildings and large structures. Reinforced concrete is concrete with steel rods under tension through it to provide additional strength.

Solubility of group II sulfates and hydroxides

- The solubility of the group II sulfates *decreases* down the group.

  $MgSO_4$ is soluble in water, but $BaSO_4$ is insoluble in water.
- The solubility of the group II hydroxides *increases* down the group.

  $Ca(OH)_2$ is partially soluble in water, but $Ba(OH)_2$ is soluble in water.

Solubility is defined as the maximum mass of solute that will dissolve in 100 g of solvent at a stated temperature.

Exam tip

To remember these trends, think about:

- magnesium reacting with dilute sulfuric acid: a colourless solution is formed so $MgSO_4$ is soluble in water
- barium chloride reacting with a solution containing sulfate ions: a white precipitate is formed — the white ppt is $BaSO_4$ so $BaSO_4$ is insoluble in water

The group II sulfates become less soluble as the group is descended and the trend is reversed for the hydroxides.

Solubility curves

- A solubility curve is a plot of solubility against temperature.
- The solvent is usually water.
- For solids the mass is measured in grams; for gases the mass may be in milligrams.
- Most solid solutes increase in solubility as temperature increases.
- Gaseous solutes decrease in solubility as solubility increases.

Table 24 gives the solubility values of the group II hydroxides and sulfates at 20°C. All values are given to three significant figures.

You can see that the solubility of the group II sulfates decreases and the solubility of the group II hydroxides increases.

Table 25 shows the solubility values of strontium hydroxide, $Sr(OH)_2.8H_2O$, at different temperatures.

These values may be plotted on a graph called a solubility curve (Figure 87).

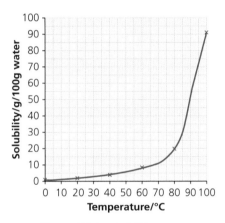

Figure 87 A solubility curve

- The solubility curve allows calculation of solubility values between given temperatures.
- When a saturated solution is cooled from a higher temperature to a lower temperature some of the solid solute recrystallises out of solution.
- The mass of solid which recrystallises depends on the change in temperature and also on the mass of solvent in the solution.

Table 24 Solubility of group II hydroxides and sulfates

Substance	Solubility (g/100g water) at 20°C
$MgSO_4$	35.1
$CaSO_4$	0.255
$SrSO_4$	0.0132
$BaSO_4$	0.000244
$Mg(OH)_2$	0.000963
$Ca(OH)_2$	0.173
$Sr(OH)_2$	1.77
$Ba(OH)_2$	3.89

Table 25 Solubility of strontium hydroxide at different temperatures

Temperature (°C)	Solubility (g/100g water) at 20°C
0	0.91
20	1.77
40	3.95
60	8.42
80	20.2
100	91.2

Knowledge check 29

What is meant by the term solubility?

Worked example

Calculate the mass of strontium hydroxide which recrystallises when a saturated solution containing 50 g of water is cooled from 90°C to 70°C.

The solubility of strontium hydroxide at 90°C is 52 g/100 g water.

The solubility of strontium hydroxide at 70°C is 11 g/100 g water.

➡

The mass which would recrystallise from $100\,g$ of water when a saturated solution is cooled from 90°C to 70°C is $52 - 11 = 31\,g$.

The mass which would recrystallise from $50\,g$ of water when a saturated solution is cooled from 90°C to 70°C is $\dfrac{31}{2} = 15.5\,g$

Summary

- Going down group II the trends in properties are:
  - reactivity increases
  - atomic radius increases
  - first ionisation energy decreases
  - thermal stability of the carbonates increases
  - solubility of the sulfates decreases
  - solubility of the hydroxides increases.
- The thermal stability of the group II carbonates can be explained on the basis of the size of the group II ion. The larger charge density of the Mg^{2+} ion polarises and destabilises the carbonate ion. Less heat is required to thermally decompose $MgCO_3$ compared to the other group II carbonates.
- Solubility is the maximum mass of solute which will dissolve in $100\,g$ of solvent at a stated temperature.

Questions & Answers

The AS2 examination is 1 hour 30 minutes in length and consists of ten multiple-choice questions (each worth 1 mark) and several structured questions which vary in length. The structured questions make up the remaining 80 marks giving 90 marks in total for the paper. For each multiple-choice question there is one correct answer and at least one clear distractor. The mark allocations for the structured questions vary but a general rule is that each error loses you a mark.

About this section

This section contains a mix of multiple-choice and structured questions similar to those you can expect to find in the A-level papers.

Each question is followed by brief guidance on how to approach the question and also where you could make errors (shown by 🅔). Answers to some questions are then followed by comments. These are preceded by 🅔. Try the questions first to see how you get on and then check the answers and comments.

General tips

- Be accurate with your learning at this level — examiners will penalise incorrect wording.
- Always follow calculations through to the end, even if you feel you have made a mistake — there are marks for the correct method even if the final answer is incorrect.
- If you round a number from your calculator, use the rounded answer to continue the calculation.
- Always attempt to answer a multiple-choice question even if it is a guess (you have a 25% chance of getting it right).
- If a quantity has units, always include them (mol is often left out).

The uniform mark you receive for AS2 will be out of 96. Both AS1 and AS2 are awarded out of 96 and the AS3 (practical examination examining practical and planning from AS1 and AS2) is awarded out of 48 uniform marks, giving a possible total of 240 for AS chemistry. AS chemistry makes up 40% of the overall A-level.

Formulae and amounts of substance

Question 1

An organic compound consists of 49.3% carbon, 6.9% hydrogen and 43.8% oxygen. The relative molecular mass of the compound is 146. Which one of the following is the molecular formula of the compound?

A $C_5H_6O_5$ **C** $C_7H_{14}O_3$

B $C_6H_{10}O_4$ **D** $C_8H_{18}O_2$ (1 mark)

ⓔ All the options have a relative molecular mass of 146 — otherwise the question would be too easy. Common mistakes would be to use atomic number instead of RAM or using diatomic RMMs for hydrogen and oxygen. The empirical formula you determine allows you to calculate the molecular formula using the RMM.

> **Student answer**
>
> moles of C $= \dfrac{49.3}{12} = 4.108\,mol$
>
> moles of H $= \dfrac{6.9}{1} = 6.9\,mol$
>
> moles of O $= \dfrac{43.8}{16} = 2.738\,mol$
>
> simplest ratio $= 1.5 : 2.5 : 1 = 3 : 5 : 2$
>
> RMM of $C_3H_5O_2 = 73$
>
> $146/73 = 2$ so $2 \times C_3H_5O_2 = C_6H_{10}O_4$
>
> Answer is B ✓

ⓔ Because the percentages are given to 1 decimal place and because of rounding in the number of moles, the ratio may be slightly out. The simplest ratio works out to be $1.5003 : 2.5201 : 1$ (where the lowest number of moles (2.738) is taken to be 1). This is close to $1.5 : 2.5 : 1$. You have to use your judgement as the answers will be close to a whole number, to 0.5 or at worst to 0.333. Work out the moles to at least three decimal places and the ratio will be more accurate.

Question 2

5.1 g of iodoethane were prepared by refluxing 5.0 cm³ of ethanol with 23.2 g of phosphorus(ɪɪɪ) iodide.

$$3C_2H_5OH + PI_3 \rightarrow 3C_2H_5I + H_3PO_3$$

The density of ethanol is 0.79 g cm⁻³.

(a) **(i)** What mass of ethanol was used?   (1 mark)

        (ii) Calculate the number of moles of ethanol used.   (1 mark)

        (iii) Calculate the number of moles of phosphorus(ɪɪɪ) iodide used.   (1 mark)

        (iv) Calculate the maximum number of moles of iodoethane that could be formed.   (1 mark)

    (v) Calculate the maximum mass of iodoethane that could be formed.   (1 mark)

    (vi) Calculate the percentage yield. Give your answer to 1 decimal place.   (1 mark)

(b) Calculate the atom economy of this reaction given that iodoethane is the desired product. Give your answer to 1 decimal place.   (1 mark)

ⓔ A common question is a percentage yield calculation in which one of the reactants is in excess and one is limiting. One reactant could be a liquid, in which case you will be given a density. You should work out the number of moles of each reactant present and then use the ratio in the balanced equation to determine which reactant is in excess and which is limiting. The number of moles of the limiting reactant is used to determine the maximum number of moles (theoretical yield in moles) of the product. Percentage yield is the mass of product obtained (actual yield) divided by the theoretical yield, multiplied by 100. It can also be worked out from number of moles of the product. Work throughout the question to 3 significant figures and then give the answer to any given level of accuracy.

Student answer

(a) (i) mass of ethanol = volume × density = $5.0 \times 0.79 = 3.95\,g$ ✓

    (ii) moles of ethanol = $\dfrac{3.95}{46} = 0.0859\,mol$ ✓

    (iii) moles of phosphorus(III) iodide = $\dfrac{23.2}{412} = 0.0563\,mol$ ✓

    (iv) maximum moles of iodoethane = $0.0859\,mol$ ✓

        (ethanol is limiting reactant and 3 : 3 ratio of ethanol : iodoethane)

    (v) maximum mass of iodoethane = $0.0859 \times 156 = 13.4\,g$ ✓

    (vi) percentage yield = $\dfrac{\text{actual yield}}{\text{theoretical yield}} \times 100 = \dfrac{5.1}{13.4} \times 100 = 38.1\%$ ✓

ⓔ Even if you cannot work out which is the limiting reactant, you should carry on with the calculation using a number for part (iv) because you would then be awarded marks for the correct method in parts (v) and (vi).

(b) atom economy = $\dfrac{\text{mass of desired product}}{\text{total mass of products}} \times 100$

         = $\dfrac{3 \times 156}{(3 \times 156) + 82} \times 100 = 85.1\%$ ✓

Question 3

$15\,cm^3$ of methane is burnt in $100\,cm^3$ of oxygen (an excess). What volume of gas remains after combustion and cooling to room temperature?

A $70\,cm^3$ **C** $100\,cm^3$

B $85\,cm^3$ **D** $115\,cm^3$ (1 mark)

> **Student answer**
>
> Answer is B ✓

ⓔ The equation for the combustion of methane is: $CH_4 + 2O_2 \rightarrow CO_2 + 2H_2O$. Using Avogadro's law $15\,cm^3$ of CH_4 reacts with $30\,cm^3$ of O_2 to form $15\,cm^3$ of CO_2 and $30\,cm^3$ of H_2O. Cooling to room temperature means that the water condenses and is not part of the gas volume. All the CH_4 is used up; $30\,cm^3$ of the O_2 is used up (leaving $70\,cm^3$) and $15\,cm^3$ of CO_2 are formed. The total gas volume after combustion (the excess O_2 and the CO_2) and cooling to room temperature is $85\,cm^3$.

Alkanes and alkenes

Questions on nomenclature and isomerism will be assessed throughout organic chemistry.

Question 1

The correct systematic name for the organic compound with the structure shown is:

A 1,1,3-trimethylpentane

B 3,5,5-trimethylpentane

C 2,4-dimethylhexane

D 3,5-dimethylhexane

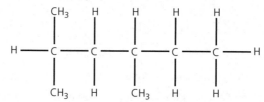

(1 mark)

> **Student answer**
>
> Answer is C ✓

ⓔ This type of question tests your ability to count the longest carbon chain and number it correctly. There can never be a 1-methyl on a substituted alkane as this methyl group is part of the main chain. If there is no functional group to define the numbering, number the carbon chain from the end that gives the lowest locant numbers for any substituent groups.

Question 2

Which one of the following is a product of the reaction of chlorine with fluoromethane?

A 1,1-difluoroethane   **C** 1,2-dichloroethane

B 1-chloro-2-fluoroethane   **D** 1,2-diflurorethane

(1 mark)

> **Student answer**
>
> Answer is D ✓

ⓔ The key to this question is to check the radical formed in the propagation step. It is $CH_2F\bullet$. Two of these combine in a termination step to form CH_2FCH_2F. This means the substituted ethane molecule has two fluoro groups, one bonded to each of the carbon atoms. Draw out the mechanism as shown below to make sure you know what is happening.

Initiation: $Cl_2 \rightarrow 2Cl\bullet$

Propagation: $CH_3F + Cl\bullet \rightarrow CH_2F\bullet + HCl$

                     $CH_2F\bullet + Cl_2 \rightarrow CH_2FCl + Cl\bullet$

Termination: $2Cl\bullet \rightarrow Cl_2$

                     $CH_2F\bullet + Cl\bullet \rightarrow CH_2FCl$

                     $2CH_2F\bullet \rightarrow CH_2FCH_2F$

Question 3

2-bromobut-2-ene, $CH_3CBr{=}CHCH_3$, is a bromo derivative of but-2-ene. It exists as *E–Z* isomers.

(a) Draw and label the structures of the *E* and *Z* isomers of 2-bromobut-2-ene.   (3 marks)

(b) 2-bromobut-2-ene reacts with hydrogen.
   (i) Describe the conditions, including any catalyst required, for this reaction.   (2 marks)
   (ii) Write an equation for the reaction of 2-bromobut-2-ene with hydrogen.   (1 mark)
   (iii) Name the product of the reaction.   (1 mark)

Student answer

(a)

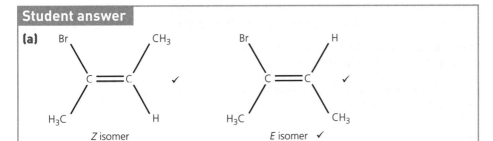

Z isomer   *E* isomer ✓

ⓔ There is 1 mark for each structure and 1 mark for correct labelling. When drawing the structures of *E–Z* or *cis–trans* isomers it is best to use the format shown above and to add the groups on. Simplify each group if necessary. The higher priority group on the left-hand carbon is Br. CH_3 has higher priority on the right-hand carbon, so in the first structure the two higher priority groups are on the same side of the axis of the $C{=}C$ bond (*Z*).

(b) (i) Nickel catalyst ✓
      Finely divided catalyst/180°C ✓

ⓔ Being able to recall information such as reagents, catalysts, types of reaction, conditions and mechanisms is important in organic chemistry. Make sure you know the details necessary for all types of reactions.

   (ii) $CH_3CBr{=}CHCH_3 + H_2 \rightarrow CH_3CHBrCH_2CH_3$ ✓

ⓔ When asked for 'an equation' in organic chemistry you can write a structural or symbol equation such as the one shown above. It is good practice to use the style of any formula given in the stem of the question. If a structural equation is asked for, draw out all organic structures showing all bonds.

> **(iii)** 2-bromobutane ✓

ⓔ Naming an organic compound based on a structural or condensed structural formula is an important skill. If you are unsure with a condensed structural formula, sketch it out.

Question 4

The reactions of methane and ethene are different. Methane reacts with chlorine in the presence of ultraviolet light and ethene reacts readily with hydrogen bromide.

(a) (i) **Name the mechanism by which methane reacts with chlorine in the presence of ultraviolet light.** (1 mark)

(ii) **Write one termination reaction from this mechanism.** (1 mark)

(b) (i) **Draw a flow scheme to show the mechanism of the reaction of ethene with hydrogen bromide. Curly arrows should be included.** (4 marks)

(ii) **Name the mechanism for the reaction between ethene and hydrogen bromide.** (1 mark)

(c) **Describe and explain the reactions of ethene compared with those of methane in terms of the types of covalent bond present.** (4 marks)

ⓔ The reactions of organic compounds within a homologous series are similar. If a molecule has a particular functional group, then this group will react in the same way with the same reactant under the same conditions. Comparing reactivity of alkanes and alkenes is a common question. Alkanes are much less reactive than alkenes and undergo substitution reactions, whereas the $C=C$ bond in alkenes allows them to undergo addition reactions. Mechanisms in AS2 are very important and you should practise drawing them. There are four to learn and you should also remember the type of reaction to which they apply. For example, for any alkene where the $C=C$ bond is converted to a $C-C$ bond, the covalent π-bond is broken and the reaction mechanism is an electrophilic addition. The most common example is ethene reacting with hydrogen bromide, but it can apply to any alkene undergoing an addition reaction.

Student answer

(a) (i) Free radical (photochemical) substitution ✓

(ii) Any *one* reaction from:

$Cl\bullet + Cl\bullet \rightarrow Cl_2$

$CH_3\bullet + Cl\bullet \rightarrow CH_3Cl$

$CH_3\bullet + CH_3\bullet \rightarrow C_2H_6$

(b)

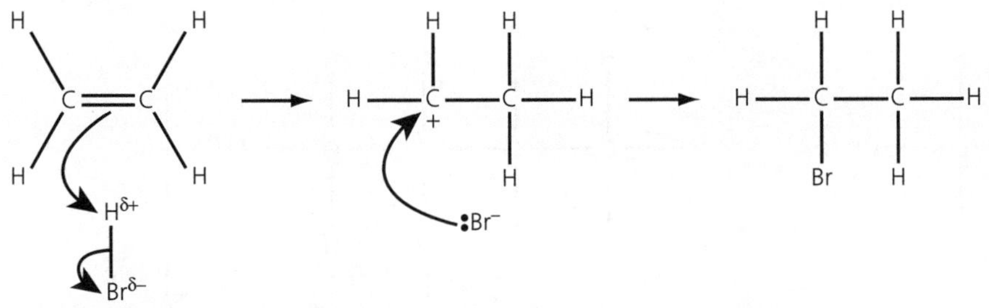

   (i) Mechanism correct ✓✓✓

      Curly arrows correct ✓

   (ii) Electrophilic addition ✓

(c) Ethene undergoes addition reactions ✓

   π-bond present in ethene but only sigma bonds present in methane ✓

   Methane undergoes substitution reactions ✓

   π-bond weaker/attacked by electrophiles ✓

Halogenoalkanes

Question 1

The flow scheme below shows some of the reactions of 1-bromobutane. Draw the structure of the organic products from the reactions shown.

(4 marks)

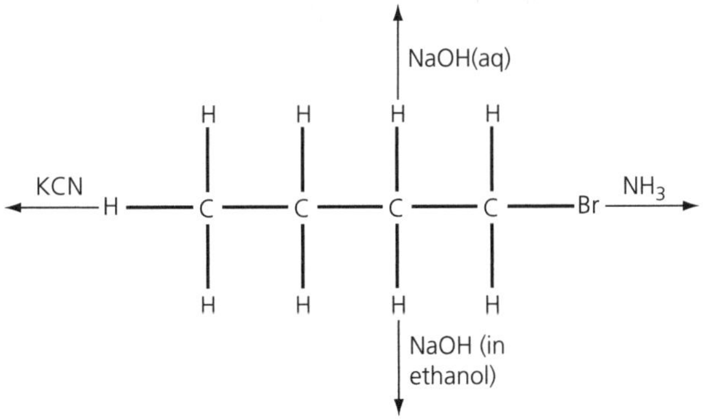

ℯ This type of question, involving different reactions of a particular homologous series, is common. Most reactions of halogenoalkanes are substitution reactions in which the halogen atom is substituted by a different group (OH, NH_2 and CN in these examples). With hydroxide ions dissolved in ethanol, an elimination reaction occurs in which the hydrogen halide (HBr in this case) is eliminated to form an alkene.

Student answer

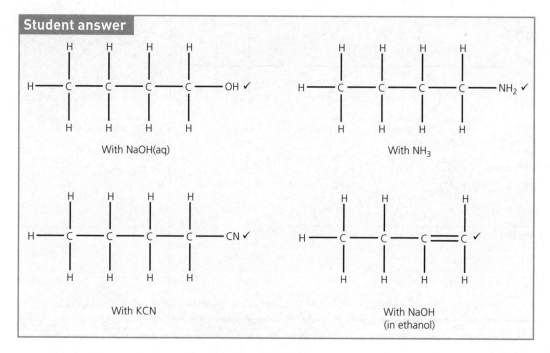

With NaOH(aq)

With NH$_3$

With KCN

With NaOH
(in ethanol)

Question 2

The mechanism for the reaction between aqueous hydroxide ions and 1-bromopropane is:

A   electrophilic addition

C   nucleophilic addition

B   electrophilic substitution

D   nucleophilic substitution   (1 mark)

ⓔ Halogenoalkanes react with aqueous sodium hydroxide by nucleophilic substitution. It is important that you can recognise the compound as a halogenoalkane and relate its reaction with aqueous hydroxide ions to the type of mechanism.

Student answer

Answer is D ✓

Question 3

1-bromobutane may be prepared using a mixture of butan-1-ol, water, sodium bromide and concentrated sulfuric acid. The mixture is heated under reflux. The impure 1-bromobutane is distilled off and washed with sodium carbonate solution before being mixed with anhydrous sodium sulfate.

(a) Sodium bromide and concentrated sulfuric acid produce hydrogen bromide in situ.

  (i)   What do you understand by the term 'in situ'?   (1 mark)

  (ii) Write an equation for the production of hydrogen bromide in the reaction mixture.   (1 mark)

(b) What is meant by 'heating under reflux'?   (1 mark)

(c) Sodium carbonate solution is mixed with the impure organic compound in a separating funnel.

   (i) Explain why the mixture is washed with sodium carbonate solution. (1 mark)

   (ii) After washing, the mixture separates into two layers. Explain how you would identify the organic layer. (2 marks)

(d) What is the purpose of the anhydrous sodium sulfate? (1 mark)

ⓔ Questions on experiments are common in AS2 papers and in the AS practical exam. You need to understand the purpose of each process. The actual experiment may not be the one you carried out in school, but the processes within it to produce a pure organic liquid will be similar.

> **Student answer**
>
> (a) (i) Produced in the reaction mixture ✓
>
>   (ii) $NaBr + H_2SO_4 \rightarrow NaHSO_4 + HBr$ ✓
>
> (b) Repeated boiling and condensing of a reaction mixture ✓
>
> (c) (i) To remove acid ✓
>
>   (ii) Add water; the layer that increases in volume is the aqueous layer ✓
>   The other layer is organic ✓
>
> (d) To remove water/drying agent ✓

Alcohols and infrared spectroscopy

Question 1

Which one of the following alcohols, all with the formula C_4H_9OH, would not react with acidified potassium dichromate(vi) solution?

A   2-methylpropan-1-ol   C   butan-1-ol

B   2-methylpropan-2-ol   D   butan-2-ol   (1 mark)

ⓔ This question could have simply asked 'Which one of the following would not undergo (mild) oxidation?' However, you are expected to recognise the reagent and result of an oxidation test. The structure of the alcohols could have been given, which would have made it easier to identify the one that is a tertiary alcohol. This could also be set as a structured question asking for all the isomers of C_4H_9OH and their names, together with details of how to carry out an oxidation test, and which of the alcohols with formula C_4H_9OH would **not** give a positive test.

> **Student answer**
>
> Answer is B ✓

ⓔ B is the only one that is a tertiary alcohol.

Question 2

Explain how ethanol absorbs infrared radiation. (2 marks)

ⓔ Organic molecules absorb infrared radiation owing to the presence of covalent bonds within the molecule, which can bend and stretch. The bending and stretching of these covalent bonds may be referred to as molecular vibrations. You must be able to recognise the source of absorption when answering either a multiple choice or a structured question.

> **Student answer**
>
> Molecular vibrations ✓ due to bending and stretching of the covalent bonds in ethanol. ✓

Question 3

The scheme shows some of the reactions of ethanol.

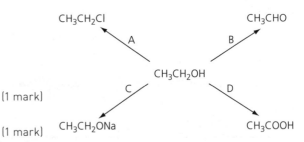

(a) Phosphorus pentachloride can be used to carry out reaction A.

　(i) Write an equation for the reaction of ethanol with phosphorus pentachloride. (1 mark)

　(ii) Name one other reagent that could be used for reaction A. (1 mark)

(b) Reactions B and D are both examples of oxidation reactions.

　(i) Name an oxidising agent that could be used for these reactions. (1 mark)

　(ii) State how the conditions for the oxidation reactions in B and D are different. (2 marks)

(c) **(i)** Name the organic substance, CH_3CH_2ONa, formed in reaction C. (1 mark)

　(ii) Draw the structure of CH_3CH_2ONa. (1 mark)

ⓔ In this type of question, it is the reactions of the OH group in alcohols that are being examined. You need to know the reagent(s) and conditions for a particular reaction, as well as being able to recognise the product formed and to write an equation for the reaction. Questions may be set on compounds with two or more of the same functional group. 'Excess reagent' indicates that all of the identical functional groups react (in the same way).

> **Student answer**
>
> **(a) (i)** $CH_3CH_2OH + PCl_5 \rightarrow CH_3CH_2Cl + POCl_3 + HCl$ ✓

ⓔ All reactions of OH groups with PCl_5 occur in the same way. The OH group is replaced by a Cl atom. The products are the halogenoalkane, $POCl_3$, and HCl.

(ii) HCl ✓

(b) (i) Acidified (potassium) dichromate solution ✓

(ii) In B, aldehyde/CH_3CHO formed on heating under distillation ✓

In D, acid/CH_3COOH formed on heating under reflux ✓

ⓔ The oxidation of primary alcohols such as ethanol occurs in two stages. The primary alcohol is first oxidised to the aldehyde, which is then further oxidised to a carboxylic acid. The oxidation to the carboxylic acid requires heating under reflux. Heating while distilling removes the aldehyde as soon as it is formed. Be aware of conditions in organic reactions, as these are important.

(c) (i) Sodium ethoxide ✓

(ii)

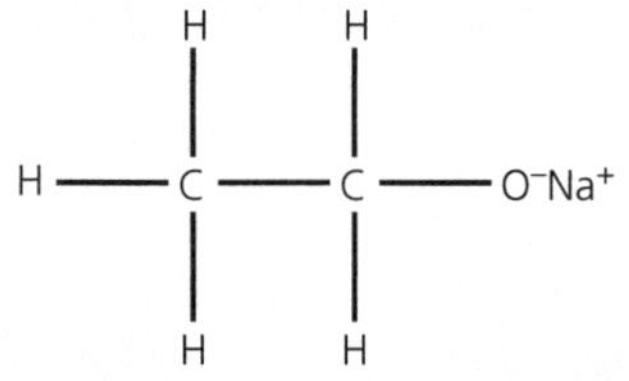

ⓔ Watch out for the structure of sodium ethoxide. Make sure you show the charge between the O^- and the Na^+. A common mistake is to draw a covalent bond. This is also true for salts of carboxylic acids.

Energetics

Question 1

The standard enthalpy change of combustion of propane, C_3H_8, is $-2200\,kJ\,mol^{-1}$. A sample of 0.880 g of propane was burnt completely in air. The heat produced was used to heat 200 g of water. Calculate the maximum temperature change observed in the water given that the specific heat capacity of water is $4.2\,J\,g^{-1}\,K^{-1}$. Give your answer to 1 decimal place. (3 marks)

ⓔ In this question you are calculating ΔT rather than using ΔT to calculate the standard enthalpy change of combustion. The same method is used but reversed.

Student answer

2200 kJ (2 200 000 J) of energy released when 1 mol of C_3H_8 is burnt

moles of propane burnt $= \dfrac{0.88}{44} = 0.02\,mol$ ✓

energy released when 0.02 mol of propane are burnt $= 0.02 \times 2\,200\,000 = 44\,000\,J$ ✓

$q = mc\Delta T$

$44\,000 = 200 \times 4.2 \times \Delta T$

$\Delta T = \dfrac{44\,000}{200 \times 4.2} = \dfrac{44\,000}{840} = 52.4°C$ ✓

Questions & Answers

ⓔ Be careful with calculations like thise. If you key $44\,000 \div 200 \times 4.2$ into your calculator, it will divide $44\,000$ by 200 and then multiply the answer by 4.2. It is good practice to either calculate the numerator (top) and denominator (bottom) separately and then do the division, or put the whole of the bottom line in brackets in your calculator.

Question 2

Sodium nitrate(v) thermally decomposes forming sodium nitrate(ɪɪɪ) and oxygen gas according to the equation:

$$2NaNO_3(s) \rightarrow 2NaNO_2(s) + O_2(g)$$

The standard enthalpy changes of formation are:

- **sodium nitrate(v), $NaNO_3(s)$** $\Delta_fH^{\ominus} = -466.7\,kJ\,mol^{-1}$
- **sodium nitrate(ɪɪɪ), $NaNO_2(s)$** $\Delta_fH^{\ominus} = -359.4\,kJ\,mol^{-1}$

Calculate the standard enthalpy change of reaction for the thermal decomposition of sodium nitrate(v).   (2 marks)

ⓔ Most enthalpy change questions focus on organic chemistry, but can also be applied to an inorganic reaction. Standard enthalpy changes of reaction are based on the equation as written and do not need to be 'per mole' of any substance.

Student answer

Solution by equations:

Main equation:

$$2NaNO_3(s) \rightarrow 2NaNO_2(s) + O_2(g)$$

Equations for given enthalpy changes:

$$Na(s) + \tfrac{1}{2}N_2(g) + 1\tfrac{1}{2}O_2(g) \rightarrow NaNO_3(s) \quad \Delta_fH^{\ominus} = -466.7\,kJ\,mol^{-1}$$

2 mol of $NaNO_3(s)$ react in the main equation, so the equation above will be reversed and multiplied by 2:

$$Na(s) + \tfrac{1}{2}N_2(g) + O_2(g) \rightarrow NaNO_2(s) \quad \Delta_fH^{\ominus} = -359.4\,kJ\,mol^{-1}$$

2 mol of $NaNO_2(s)$ are formed in the main equation, so the equation above will be multiplied by 2:

$$2NaNO_3(s) \rightarrow 2Na(s) + N_2(g) + 3O_2(g) \qquad 2(+466.7)$$
$$2Na(s) + N_2(g) + 2O_2(g) \rightarrow 2NaNO_2(s) \qquad 2(-359.4) ✓$$

$2NaNO_3(s) + 2Na(s)$ $+ N_2(g) + 2O_2(g)$	$\rightarrow$	$2Na(s) + N_2(g) +$ $3O_2(g) + 2NaNO_2(s)$	$2(+466.7)+$ $2(-359.4)$
$2NaNO_3(s)$	$\rightarrow$	$2NaNO_2(s) + O_2(g)$	$+214.6$✓ kJ

Solution by Hess's law diagram

$\Delta H^{\ominus} = -2(-466.7) + 2(-359.4)$ ✓

$= +933.4 - 718.8 = + 214.6$ ✓ kJ

Or more simply:

$\Delta H = \Sigma\Delta_f H(\text{products}) - \Sigma\Delta_f H(\text{reactants})$

$\Delta H = 2(-395.4) - 2(-466.7)$ ✓ $= +214.6$ ✓ kJ

$$2NaNO_3(s) \rightarrow 2NaNO_2(s) + O_2(g)$$

$2(-466.7) \quad 2(-359.4) \quad 0$

$2Na(s) \quad N_2(g) \quad 3O_2(g)$

e This reaction is endothermic overall, as you would expect a thermal decomposition to be. Remember to include the sign (+ or −) in front of the enthalpy change value — leaving this out can cost you a mark.

Question 3

Propene gas burns in oxygen according to the equation:

$$2C_3H_6(g) + 9O_2(g) \rightarrow 6CO_2(g) + 6H_2O(l)$$

The bond enthalpy values are given in the table.

Which one of the following is the standard enthalpy change of combustion of propene?

Bond	Average bond enthalpy/$kJ\,mol^{-1}$
C—H	413
C—C	346
C=C	611
O—H	464
O=O	496
C=O	803

A $-1670\,kJ\,mol^{-1}$

B $-1935\,kJ\,mol^{-1}$

C $-3340\,kJ\,mol^{-1}$

D $-3870\,kJ\,mol^{-1}$

(1 mark)

Student answer

Answer is B ✓

e The main issues with this style of question are using the bond enthalpy values correctly and then dividing by 2, so that the enthalpy change is calculated for 1 mol of propene as opposed to the 2 mol in the equation. The total enthalpy required to break the bonds, based on the main equation given, is: $2 \times 346 + 2 \times 611 + 12 \times 413 + 9 \times 496 = 11334\,kJ$. The energy released when bonds form, based on the main equation, is: $12 \times 803 + 12 \times 464 = 15204\,kJ$. The enthalpy change in the equation is $+11334 - 15204 = -3870\,kJ$. However this is for 2 mol of propene, so the correct answer is $-3870/2 = -1935\,kJ\,mol^{-1}$. -3870 is a common incorrect answer.

Kinetics and equilibrium

Question 1

In the Haber process nitrogen reacts with hydrogen according to the equation:

$$N_2(g) + 3H_2(g) \rightleftharpoons 2NH_3(g) \quad \Delta H = -92.2\,kJ$$

The reaction is carried out at a temperature of 450°C and 200 atm pressure. An iron catalyst is used.

Questions & Answers

(a) The diagram shows the distribution of molecular energies in the reaction mixture at 450°C.

(i) Sketch on the diagram the distribution of molecular energies at 550°C. (1 mark)

(ii) Using the diagram, explain why the rate of reaction would be faster at 550°C. (2 marks)

(iii) Explain how the yield of ammonia is affected by increasing the temperature to 550°C. (2 marks)

(b) Explain why a high pressure is used. (2 marks)

(c) With reference to the diagram, explain how a catalyst increases the rate of reaction. (2 marks)

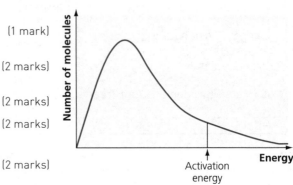

Student answer

(a) (i)

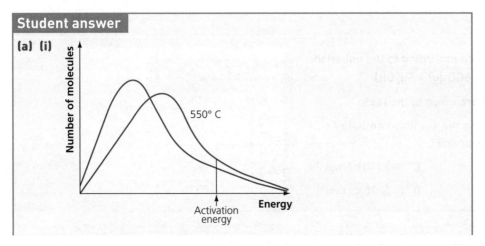

ⓔ The sketch for 550°C should be lower and more to the right.

> (ii) There are more molecules above the activation energy ✓, so there are more successful collisions. ✓

ⓔ Try to keep the area under the curves roughly the same. At a higher temperature the curve should be more spread out with a lower peak. This is a common question and you should understand the link between the shape of the curve and the rate of reaction in terms of the activation energy.

> (iii) The reaction is exothermic, so increasing the temperature moves the position of equilibrium in the direction of the reverse reaction/position of equilibrium moves to the left. ✓
> The yield of NH_3 decreases. ✓

ⓔ A question on yield always requires an answer as to whether the yield increases or decreases. Make sure you include this in your answer — just stating that the position of equilibrium moves to the left is not enough.

(b) Fewer moles of gas/4 mol of gas on left and 2 mol of gas on right *and* the position of equilibrium moves to the right to reduce volume/minimise effect of increased pressure. ✓

Higher rate of reaction at increased pressure. ✓

ⓔ Questions about the effects of temperature, pressure and the presence of a catalyst on rate and on equilibrium are common. Both the number of gas moles on each side and the equilibrium position moving to the right are required for the first mark. This is common at AS, so give as full an answer as possible to ensure you do not lose marks unnecessarily.

(c) (i) A catalyst provides an alternative reaction pathway of lower activation energy. ✓ There are more molecules with energy greater than activation energy/more successful collisions. ✓

ⓔ A catalyst does not affect the position of equilibrium but increases the rate by providing an alternative reaction pathway of lower activation energy. This means that equilibrium is reached more quickly. The second mark is for referring to the number of molecules with energy greater than the activation energy.

Question 2

The reaction of hydrogen with iodine is an equilibrium.
$$H_2(g) + I_2(g) \rightleftharpoons 2HI(g) \qquad \Delta H = -10.4\,kJ$$

Which one of the following would move the position of equilibrium to the right?

A   adding hydrogen iodide

C   decreasing the pressure

B   decreasing the temperature

D   adding a catalyst   (1 mark)

Student answer
Answer is B ✓

ⓔ Adding hydrogen iodide would move the position of equilibrium to the left. As there is an equal number of moles of gas on each side of the equation, a change in pressure has no effect on the position of equilibrium. A catalyst does not affect the position of equilibrium, it increases the rate of reaction.

Question 3

For the homogeneous gaseous equilibrium:
$$4NH_3(g) + 3O_2(g) \rightleftharpoons 2N_2(g) + 6H_2O(g)$$

(a) Write an expression for K_c for this reaction.   (1 mark)

(b) State the units of K_c.   (1 mark)

(c) The reaction is exothermic. Explain what effect, if any, an increase in temperature would have on the value of K_c and the position of equilibrium.   (3 marks)

(d) Explain what effect, if any, an increase in pressure would have on the position of equilibrium and the value of K_c.   (4 marks)

(a) $K_c = \dfrac{[N_2]^2[H_2O]^6}{[NH_3]^4[O_2]^3}$ ✓

(b) $mol\,dm^{-3}$ ✓

(c) K_c decreases ✓

Position of equilibrium moves from right to left ✓

To absorb the heat ✓

(d) K_c does not change ✓

Position of equilibrium moves from right to left ✓

Smaller gas volume on left-hand side/8 mol of gas on right and 7 mol of gas on left ✓

Position of equilibrium moves to minimise the effect of the increase in pressure ✓

ⓔ The units are calculated from the K_c expression: $(mol\,dm^{-3})^8$ on the top line and $(mol\,dm^{-3})^7$ on the bottom line, which gives $mol\,dm^{-3}$. As the forward reaction is exothermic, an increase in temperature moves the position of equilibrium from right to left, because the equilibrium will move to minimise the effect of the increase in temperature so it will absorb the heat. K_c decreases as temperature increases as the concentration of the products (on the top line) decreases. An increase in pressure will move the position of equilibrium to the side that has the lower gas volume. This is the left-hand side. Count the number of moles of gas on each side and state this in your answer. K_c is only affected by a change in temperature, so a change in pressure may move the position of equilibrium but will not change the value of K_c.

Group II elements and their compounds

Question 1

Strontium carbonate undergoes thermal decomposition.

(a) Write an equation for the thermal decomposition of strontium carbonate.

(1 mark)

(b) Strontium carbonate is more thermally stable than magnesium carbonate. Explain this difference.

(2 marks)

ⓔ This is a standard style of question on group II chemistry. You should be able to write equations for all the reactions, many of which you should have met at GCSE. Learn the differences in the thermal stability of the carbonates and how to explain them.

(a) $SrCO_3 \rightarrow SrO + CO_2$ ✓

(b) Sr^{2+} larger/Mg^{2+} smaller ✓

Sr^{2+} lower charge density so less polarising/destabilising effect on carbonate ion ✓

Question 2

Which of the following shows the correct order of increasing solubility of the compounds listed?

A barium sulfate, calcium sulfate, magnesium sulfate

B calcium sulfate, magnesium sulfate, barium sulfate

C magnesium sulfate, calcium sulfate, barium sulfate

D barium sulfate, magnesium sulfate, calcium sulfate   (1 mark)

> Answer is A ✓

🅔 The solubility of the sulfates decreases down the group. The correct order of increasing solubility is barium sulfate, (strontium sulfate), calcium sulfate and magnesium sulfate.

Question 3

Calcium carbonate decomposes on heating to calcium oxide. Calcium oxide reacts with water to form calcium hydroxide.

(a) What common name is used for calcium oxide?   (1 mark)

(b) Write an equation for the reaction of calcium oxide with water.   (1 mark)

(c) What is observed when calcium oxide reacts with water?   (2 marks)

(d) Explain why calcium oxide is described as a base.   (1 mark)

Student answer

(a) Quicklime ✓

(b) $CaO + H_2O \rightarrow Ca(OH)_2$ ✓

(c) Any two from: hisses/releases heat/expands/crumbles ✓✓

(d) It reacts with acids to form a salt and water/accepts a proton (H^+) ✓

🅔 These reactions are very important industrially. Calcium oxide is called quicklime or sometimes just lime. When it reacts with water it forms calcium hydroxide, which is known as slaked lime (a solution of calcium hydroxide is called limewater).

Knowledge check answers

1 $97.5 \, cm^3$

2 76.3%

3 2-methylpentane

4 2,2-dibromobutane

5 3-methylbutanoic acid

6 4

7 chloroethane, hydrogen chloride and butane

8 electrophilic addition

9 1,2,3,4-tetrachlorobutane

10 secondary

11 $CH_3CH_2CH_2CH_2Br + KOH \rightarrow$

$\qquad CH_3CH_2CH{=\!\!=}CH_2 + KBr + H_2O$

or $C_4H_9Br + KOH \rightarrow C_4H_8 + KBr + H_2O$

12 nucleophilic substitution/S_N2

13 2-chlorobutane, phosphorus oxychloride, hydrogen chloride

14 $CH_3CH(OH)CH_2CH_3 + [O] \rightarrow CH_3COCH_2CH_3 + H_2O$

or $C_4H_9OH + [O] \rightarrow C_4H_8O + H_2O$

15 molecular vibrations/bending and stretching of covalent bonds

16 change in enthalpy

17 298K and 100kPa pressure

18 m = mass (g); c = specific heat capacity ($J\,g^{-1}\,°C^{-1}$ or $J\,g^{-1}\,K^{-1}$) and ΔT = change in temperature (°C)

19 Hess's law states that the enthalpy change for a reaction is independent of the route taken provided the initial and final conditions are the same.

20 –65 (kJ)

21 A catalyst provides an alternative reaction pathway of lower activation energy.

22 vertical axis = number of molecules; horizontal axis = energy

23 Dynamic equilibrium is a reaction system where the amounts of the reactants and the products remain constant as the rates of the forward and the reverse reactions are equal.

24 $2HBr \rightleftharpoons H_2 + Br_2$

25 iron

26 first ionisation energy decreases and atomic radius increases

27 $Sr + 2H_2O \rightarrow Sr(OH)_2 + H_2$

28 thermal stability increases

29 maximum mass of solute which will dissolve in 100g of water at a stated temperature

Note: page numbers in **bold** font indicate key terms.

A

activation energy **77**, 78, 79
addition polymerisation, alkenes 44–45
addition reactions, alkenes **38**
alcohols 57–61
alkanes 33–37
alkenes 38–45
ammonia production 84–85
atom economy **12–13**
atomic radius **86**
average bond enthalpy **75–76**
Avogadro's law 8

B

bases **88**
boiling point
   alkanes 33
   halogenoalkanes 48–49
bond enthalpies 53, 75–76
bond length **38**, 53
bromination
   with bromine water 44
   of ethene (with HBr) 41

C

C=C double bond 29–31, 38, 44
calcium compounds, uses of 90
carbocation 41–43, 56
catalysts **77**
   effect on activation energy 79
   and equilibrium position 82
   industrial reactions 85
   and reaction rate 78
catalytic converters 35
cement 90
CFCs and ozone depletion 56–57
chain isomerism 26–27
chemisorption 77
CIP priority rules 29
cis–trans (E–Z) isomers 32
combustion
   alcohols 58–59

alkanes 33–34, 35
alkenes 41
gaseous hydrocarbons 8–10
standard enthalpy 66, 68–69
compromise temperature 85
concentration
   and equilibrium position 81
   and rate of reaction 78
concrete 90
condensed structural formula 14–15
Contact process 85

D

dienes 40
distillation apparatus 50
dynamic equilibrium **80–81**

E

electrophiles **41**
electrophilic addition 41, 52
elimination reaction **54**
empirical formulae **6**, 7, 15
endothermic reaction **65**
energetics 65–77
energy conservation principle **70**
enthalpy 65
enthalpy changes (ΔH) **66–77**
enthalpy-level diagrams 65, 78
enthalpy of reaction **67**
equilibrium **80–81**, 82–85
ethanol fuel 58–59
'excess reagent', use of 45
exothermic reaction **65**
E–Z isomerism 29–31

F

formulae
   determining 6–7
   functional group isomers 28–29
   for organic molecules 14–18
   structural isomers 26
   unknown hydrocarbons 8–10
fossil fuels, pollution from 34–35
free radical photochemical substitution 36–37, 52

free rotation 29, 38
functional group isomerism 28–29
functional groups **22**
   multiple 25–26, 45
   naming molecules with 22–25

G

gas laws 8
gas volume calculations 7–10
geometric isomerism 29–32
geometric isomers **29**
greenhouse gases 34–35
ground state **63**
group II elements 86–92
   calcium compounds 90
   oxides 88–89
   reactions 86–88
   solubility 90–92
   thermal stability 89

H

Haber process 84–85
halogenoalkanes 46–56
   CFC effects on ozone 56–57
   classification 46–47
   mechanisms 55–56
   naming 19–21, 46
   physical properties 47–49
   preparation from an alcohol 49–51
   reactions 52–54
   reactions that produce 52
heating under reflux 50
Hess's law **70–74**
heterogeneous catalyst 77, **85**
heterogeneous reaction **80–81**
heterolytic fission **41**
homogeneous catalyst 77
homogeneous reaction **80–81**
homologous series **22–23**
homolytic fission **36**, 57
hydrocarbons **33**
   see also alkanes; alkenes
   formula of unknown 8–10
hydrogenation **43**
hydrolysis **52–53**, 55–56

Index

I

identification tests 62
industrial reactions 84–85
infrared spectroscopy 63–64
initiation reactions 36
isomers **26–32**, 38–40, 49

K

K_c (equilibrium constant) 82–84
kinetics 77–80

L

locant numbers 20–26, 46

M

Maxwell–Boltzmann distribution
        79–80
mechanisms 36–37, 55–56
metal oxides 88–89
miscibility **57**
molar gas volume **7**
molecular formulae 6, 14
monohalogenation of alkanes 35
monomers **44**

N

neutralisation, standard enthalpy
        66–67
    experimental determination
        67–68
nomenclature
    alkanes 19–21
    formulae 14–18
    halogenoalkanes 46
    molecules with a functional group
        22–26
nucleophiles **55**
nucleophilic substitution 55–56

O

organic identification tests 62
oxidation
    alcohols 60–61
    catalytic converters 35

oxides, group II 88–89
ozone, effect of CFCs on 56–57

P

percentage yield 11–13
pi (π) bond **38**
pollution 34–35
polmers **44**
polymerisation **44–45**
positional isomerism 26, 28
pressure
    and equilibrium position 82
    and rate of reaction 78
primary alcohols 57, 58
    (mild) oxidation of 60
primary carbocation **42**
primary halogenoalkanes **46**
    hydrolysis of 55
principle of conservation of
        energy **70**
propagation reactions 36

R

radicals **36**
reaction rate **77**
    factors affecting 77–78
reduction reactions 35
reflux **50**
reversible reactions **80–81**

S

saturated hydrocarbon **33**
s-block element **86**
secondary alcohols 57, 58
    (mild) oxidation 61
secondary carbocation **42**, 43
secondary halogenoalkanes **46**, 47
sigma (σ) bond **38**
skeletal formula 15–18
solubility **90**
solubility curves 91
spectroscopy, infrared 63–64
standard conditions **66**
standard enthalpy change **66**

standard enthalpy of combustion **66**
standard enthalpy of formation **67**
standard enthalpy of neutralisation
        66–67
state changes **65**
stereoisomerism 29–32
straight-chain alkenes 38–40
structural formula 14–15
structural isomerism 26–29
substituent groups 19, 20, 22, 23
substitution **36**
substitution reactions
    free radical photochemical 36, 52
    nucleophilic 52, 53, 54, 55–56

T

temperature
    change, enthalpy 67–69
    compromise, ammonia
        production 85
    and equilibrium constant 84
    and equilibrium position 81–82
    Maxwell–Boltzmann
        distribution 80
    and rate of reaction 78
termination reactions 36–37
tertiary alcohols 57–58, 61
tertiary carbocation **42–43**
tertiary halogenoalkanes **46**, **47**
    hydrolysis of 56
theoretical yield 10–11
thermal stability, group II
        compounds 89

U

unsaturated hydrocarbons **38**

W

waste reduction 13
wavenumber **63**

Y

yield of product 10–12, 13,
        81–82, 85